Outsourcing in Clinical Drug Development

Roy Drucker and Graham Hughes, Editors

CRC Press
Taylor & Francis Group
Boca Raton London New York

CRC Press is an imprint of the
Taylor & Francis Group, an **informa** business

CRC Press
Taylor & Francis Group
6000 Broken Sound Parkway NW, Suite 300
Boca Raton, FL 33487-2742

First issued in paperback 2019

© 2002 by Taylor & Francis Group, LLC
CRC Press is an imprint of Taylor & Francis Group, an Informa business

No claim to original U.S. Government works

ISBN-13: 978-1-57491-112-1 (hbk)
ISBN-13: 978-0-367-39593-3 (pbk)

Visit the Taylor & Francis Web site at
http://www.taylorandfrancis.com

and the CRC Press Web site at
http://www.crcpress.com

Contents

Preface

We like to think that outsourcing is a new phenomenon—but it is not. Throughout history, powerful men have bought the services of other men or organizations of men, on a temporary, if expedient, basis. The great cathedrals of Europe benefited immeasurably from the services of such groups, whereby itinerant guilds of skilled craftsmen created great works of art on a contractor-contractee basis. Over the centuries, mercenary armies have fought great wars. The concept of a large cadre of permanent employees is, in historical terms, a relatively recent one and, in even more modern times, one that is receding, as a new era of outsourcing and subcontracting is coming upon us.

Since the beginning of the Industrial Revolution, there has been a shift from a network of cottage industries to wholesale vertical integration, resulting in so-called business perfection. In the Nineteenth Century and throughout most of the Twentieth Century, uncertainties in supplies, communication, and transportation made the possession of all capabilities under one metaphorical roof a sensible and successful business strategy. However, changes to the modern world have called this strategy increasingly into question.

Raw materials (coal, iron-ore, oil) are no longer the most powerful building blocks in today's industrial society. Increasingly, power resides in the imagination and creativity of the human brain. Modern communications and information technology (IT) have provided the capability for this human brainpower to be ever more widely distributed, and widely available on demand. This means that the old idea that big is beautiful is becoming increasingly outdated—at least when it comes to the internal resources of a company. Big can still be beautiful, however, if the organization can set up global networks and exploit external resources so that its power comes from what it can achieve, not from what it owns.

I have been involved in outsourcing—as it is now called—in the pharmaceutical industry for over twenty years. I have observed that the practices involved in outsourcing have become increasingly sophisticated and the practitioners ever more resourceful and influential. It is now almost impossible to imagine that the industry could survive without outsourcing. Although this book focuses on outsourcing drug development, the industry as a whole outsources significant parts of its activities.

Indeed, it is probably fair to say that outsourcing in drug development lags in its development behind outsourcing in other fields such as discovery, sales and marketing, manufacturing, plant construction, advertising, IT, and even strategic planning.

This book was conceived some three years ago after we at Technomark examined in detail the outsourcing practices that were "best in class" in other nonpharmaceutical industries. We felt that the industry had much to learn from these other industries; however, we felt strongly that there was much expertise in outsourcing within the industry that was poorly shared and incompletely leveraged. Hence we decided, with our publishers, to compile a handbook dealing with the main issues in outsourcing.

We have been fortunate to assemble a team of talented writers who are at the forefront of outsourcing practice, either in the mainstream industry or in the outsourcing contractor companies, more commonly called contract research organizations (CROs). An executive new to outsourcing can learn immeasurably by reading the contributions of our authors; the more experienced will find many insights into the outsourcing process that they will find useful in their daily jobs.

The practice of outsourcing is, as I have implied above, in a constant state of flux. In particular, there is a lively industry-wide debate on the relative merits of true strategic outsourcing—as discussed by Jeffrey Rudolph and Joseph Tempio in Chapter Fifteen—and of tactical outsourcing as practiced by most of the industry. Whichever side of the fence a company comes down on, however, it is clear that optimal value can only be obtained from outsourcing if it is done properly. This book, we confidently expect, will enable outsourcers to do just that.

Technomark is dedicated to assisting companies, be they "big pharma," emerging life science companies, or CROs, in the creation of maximum value from outsourcing. This book is part of that dedication.

Dr Graham Hughes
Scientific Director
Technomark Consulting Services
King House, 5-11 Westbourne Grove
London W2 4UA
United Kingdom
Telephone: +44 (0) 207 229 9239
Facsimile: +44 (0) 207 792 2587
www.technomark.com
rghughes@technomark.com

Graham Hughes

Suppliers to Healthcare R&D

Over the last twenty years the contract research organization (CRO) industry has expanded from an entrepreneurial, almost cottage industry that served the unforeseen capacity shortfalls of the pharmaceutical industry, to an industry that generates more than $6 billion annually in the drug development area alone. Upward of 20 percent of all drug development budgets are spent on a wide range of suppliers from Quintiles (a company with annual revenues in excess of $1 billion per year) to small, local CROs offering specialized local services.

It is estimated that there are more than thirteen hundred CROs offering drug development services. This book concentrates on how relations with these companies can be optimized, formalized, and managed on an ongoing basis. The book is intended for the junior to middle members of the company management team who have day-to-day responsibilities in seeing their projects home. The chapters offer practical, down-to-earth advice across the spectrum of clinical drug development.

Outsourcing in the pharmaceutical industry is a recent phenomenon, but not a new one. Even in the development field, activities have been subcontracted for many years. Companies providing toxicology and analytical services have histories that predate World War II. In other fields, outsourcing has taken off more quickly, so that today, sales, marketing, advertising, distribution, and, particularly, primary and secondary manufacture are all subcontracted extensively. Other activities, such as building management, have in some cases gone down the road of facilities management, as have the more technically demanding fields of information technology (IT) and communications. Even discovery has been extensively outsourced of late: A typical portfolio of drug development candidates now contains upward of 30 percent of drugs not originally discovered by the pharmaceutical company. These drugs may have been: developed by long-term strategic drug development partners; licensed in from biotechnology companies, research institutes, or universities; or even bought from or exchanged with the company's apparent competitors.

The term "strategic partners" requires some explanation in the context of this book. Strategic drug development outsourcing (with true partnerships based on trust, shared risk, and shared rewards) rarely exists in major, research-based, pharmaceutical companies. Strategic alliances with suppliers require the commitment of both parties to a degree of customer focus, mutual alignment of objectives, and long-term vision and planning, which a pharmaceutical company may not yet be willing or ready to undertake. Indeed, not all companies have a coherent outsourcing plan in any of the disciplines of research (clinical research; regulatory toxicology and pharmacology; and chemistry, manufacturing, and controls), let alone a policy that is applied uniformly to all of them (see Figures 1.1 and 1.2).

Strategic relationships between companies, whether those companies are similar or quite disparate in size, are long-term, planned, and focused on value. Tactical relationships, on the other hand, tend to have a short-term focus, may not be well-planned, and may focus on price and delivery rather than value. No implied value judgment on the relative merits of strategic or tactical relationships between supplier and customer is intended here; indeed, successful companies will have a range of relationships. However, the contentions advanced in this book assume that well-managed tactical relationships have many of the characteristics of strategic alliances and that the evolution of a company from a tactical outsourcer to a strategic ally can be effective only if the tactical outsourcing is done well. It is this aspect of outsourcing that the following chapters address.

SUPPLIERS TO HEALTHCARE R&D

This broad overview of the supplier scene and how to enter it concentrates on the clinical development process. Figure A.2 shows the drug development process schematically. The Appendix explains the process in some detail.

Figure 1.1   Company-wide policies for outsourcing

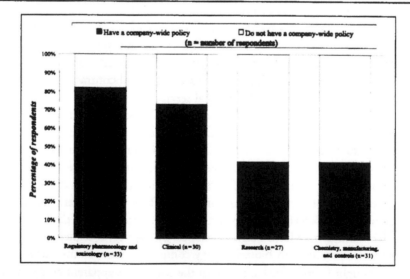

(Hughes and Lumley 1999. Reprinted with permission from Technomark Consulting Services Ltd.)

Figure 1.2   Company-wide policies in companies returning all four questionnaires

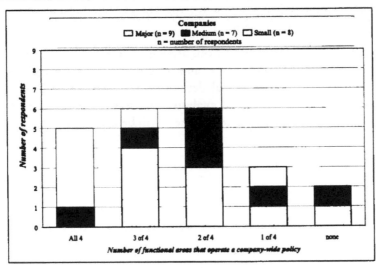

(Hughes and Lumley 1999. Reprinted with permission from Technomark Consulting Services Ltd.)

For the purpose of this book it is useful to break the clinical development process down into smaller, service-related parts:

- Phase I
- Phase II/III
- Data management and statistics

Figure 1.3   European CROs by revenue 1998–1999

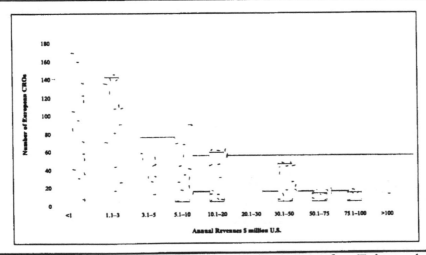

(Compiled from Technomark internal data. Printed with permission from Technomark Consulting Services Ltd.)

- Clinical trial supplies
- Regulatory affairs
- Quality assurance (QA)
- Central pathology laboratories

Later chapters discuss in detail various aspects of selecting, managing, and contracting with CROs that provide any or all of these services. We have chosen not to discuss issues such as pharmaceutical development, Phase IV studies, toxicology and animal pharmacology, chemical analysis and assay, or pharmacoeconomics in any detail in the main body of the book. These will be briefly addressed later in this chapter.

TYPES OF SUPPLIERS

CROs may be divided into a number of groups, including:

- Global multiservice providers;
- Local multiservice providers (on a national or continental basis); and
- Single-service providers (either locally or on a multicountry basis).

Over the last few years, there has been considerable consolidation of suppliers with, in particular, many of the local specialists being acquired by the global multiservice providers.

Table 1.1 shows how the current clinical market is divided among the top twenty clinical CROs.

Many previously independent Phase I units have now been acquired by the multinational CROs—particularly those in Europe (see Table 1.2). However, a substantial number of units that do the great majority of their business in Phase I remain independent (see Table 1.3 for some of the larger companies of this type).

This is not to imply that the smaller companies have all but disappeared. Figures 1.3 and 1.4 show that there are still large numbers of smaller CROs; indeed, those with annual revenues between $1 million and $1.5 million still comprise the majority of companies. Figures 1.5 and 1.6 demonstrate that this is also true in the United States.

The *Technomark Register* lists twelve companies that claim to be multinational (all included in Table 1.4). In addition to the companies whose presence is restricted either to Europe or the United States, a number of other larger companies are global, having a presence in both the United States and Europe. The majority of these companies are based in the United States.

Table 1.3 lists independent clinical research CROs with a substantial presence (more than fifty full-time employees engaged in Phase I–III clinical research, analysis, and reporting) in their own country (site management organizations [SMOs] are not listed).

The trend toward consolidation in the last few years has significantly limited the choice of a medium-sized CRO on a country-by-country basis.

Table 1.1 Leading Global Clinical CROs 1998

	Company	Country	Total Revenues ($ million U.S.)	Clinical Revenues ($ million U.S.)	Market Share
1	Quintiles Transnational	U.S.	1,188.0	617.8	17.5%
2	Covance	U.S.	731.6	351.2	9.9%
3	Parexel	U.S.	325.1	281.2	8.0%
4	PPD	U.S.	235.6	200.3	5.7%
5	Omnicare (IBAH)	U.S.	1,517.4	105.4	3.0%
6	Kendle International	U.S.	89.5	89.5	2.5%
7	UnitedHealth (Ingenix Int'l)	U.S.	17,355.0	74.3	2.1%
8	Phoenix International*	Canada	111.0	64.4	1.8%
9	PRA International	U.S.	59.8	59.8	1.7%
10	SCIREX	U.S.	60.0	59.4	1.7%
11	Applied Analytical Industries	U.S.	98.2	54.0	1.5%
12	ClinTrials Research	U.S.	89.7	51.5	1.5%
13	ICON Clinical Research	U.S.	49.8	47.3	1.3%
14	MDS*	Canada	691.0	42.9	1.2%
15	Hill Top Research	U.S.	45.0	38.3	1.1%
16	Premier Research Worldwide	U.S.	31.8	29.6	0.8%
17	ASTER-CEPHAC	France	30.0	27.0	0.8%
18	Research Triangle Institute	U.S.	134.0	24.1	0.7%
19	Pharma Bio-Research	Netherlands	29.0	23.6	0.7%
20	Statprobe	U.S.	17.0	17.0	0.5%
	Total Market			**3,531.4**	**64.0%**

*Currently merging.

(Compiled from Technomark internal data and company annual reports. Printed with permission from Technomark Consulting Services Ltd.)

Figure 1.4 Breakdown of European CROs by number of full-time employees

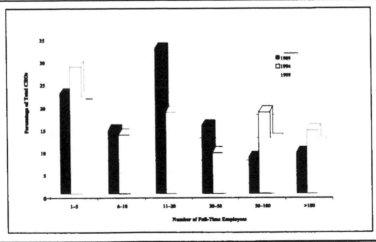

(Compiled from Technomark internal data. Printed with permission from Technomark Consulting Services Ltd.)

Table 1.2   Examples of European Phase I CROs Acquired by Multinationals

Previously Independent CRO	Multinational CRO
Guys Drug Research Unit	Quintiles
Biodesign	Quintiles
PMC	Quintiles
McKnight	Phoenix
LCRC	Pharmaco
U-Gene	Kendle

Table 1.3   CROs with More Than Fifty Full-Time Employees in Their Home Countries

Country	CRO	Foreign Subsidiaries	Member of Alliance
United States	ABT		
	Advanced Biologics		
	Advanced Biomedical		X
	Applied Logic Association		X
	Barton & Polansky		
	Boston Biostatistics		
	Cato Research	Israel, Canada	
	Clinical Trial Management Services		
	Clinicon	UK	
	GFI*		
	GloboMax	UK	
	Hill Top Research*	UK	
	Ilex	UK, Germany	
	Lineberry Research Association		
	Medex		
	MIMC	China, Germany, Canada	
	NCGS		
	Paragon BioMedical		X
	Pharmacokinetics Lab Inc.		
	PharmaNet	UK, Belgium	
	Pharma Research Corp.	UK, France, Spain	
	Premier Research	UK	
	Scirex		
Canada	Biovail		
	CroMedica	UK, Republic of South Africa, Australia	
	Endpoint Research	UK	
UK	Statprobe		
	Theradex		
	Axess		
	Hammersmith Medicines Research*		
	Inveresk Research	United States	
	Medeval*		
	Nottingham Biostatistics		
	Orion Clinical Services		

Table 1.3   continued

Country	CRO	Foreign Subsidiaries	Member of Alliance
	Phase I Clinical Trials*		
	Therapeutic Management		
	Simbec*		
France	Aster Cephac*		
	Biotrial*		
	Forenap*		
	MAPI		
	Therapharm*		
Germany	AMS Medical		
	ECRON		X
	FOCUS	UK, Eastern Europe	
	Harrison Clinical	UK, Belgium	X
	INFORM*	Czech Republic, Poland	
	IKP*		
	Staticon		
Belgium	Medisearch		
	SGS-Biopharm		
Netherlands	IMRO-TRAMARKO	France	X
	TNO	UK	
Spain	**Bio Medical Systems**		
Sweden	Clinical Data Care	Spain, United States	

*Main focus is Phase I

Table 1.4   Multinational CROs

Home Base	Clinical Company	Research	Toxicology	Laboratory	No. of Countries with Offices
U.S.	AAI	•		•	9
United Kingdom	Chiltern	•		•	7
U.S.	ClinTrials	•	•	•	11
U.S.	Covance	•	•	•	17
U.S.	IBAH	•			22
Ireland	ICON	•		•	10
U.S.	Kendle	•			6
Canada	MDS Harris	•	•	•	6
U.S.	Parexel	•			35
Canada	Phoenix	•		•	11
U.S.	PPD	•		•	9
U.S.	PRA	•			4
U.S.	Quintiles	•	•	•	31

(*Technomark Register* 1999a; 1999b; 1999c; 1999d. Reprinted with permission from Technomark Consulting Services Ltd.)

Figure 1.5   U.S. CROs by revenue 1999

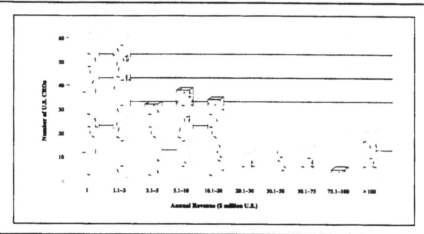

(Compiled from Technomark internal data. Printed with permission from Technomark Consulting Services Ltd.)

PHASE I SUPPLIERS

Phase I work is carried out currently in several types of locations. Many of the large pharmaceutical companies have their own internal units located within company facilities or, often, at a nearby university hospital. Others are in purpose-built units run by CROs. CROs may also lease space temporarily in hospitals and clinics.

Suppliers of Phase I services fall into two categories: commercial and not-for-profit. Many university and governmental research establishments carry out clinical pharmacology projects on behalf of the industry. Projects are undertaken by

Figure 1.6   Breakdown of U.S. CROs by number of full-time employees

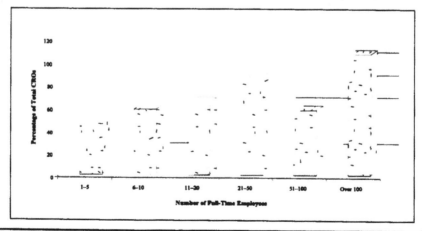

(Compiled from Technomark internal data. Printed with permission from Technomark Consulting Services Ltd.)

universities and government establishments because of their academic interest to the professor or departmental head of a governmental institution. Although the quality of the science is good, these units may not be fully aware of the current Good Clinical Practice (GCP) requirements. Being at heart noncommercial, they often let delivery times and cost control efforts slip. Increasingly, companies are moving away from placing conventional drug development projects with such institutions.

Commercial Phase I units may themselves be located inside hospitals (by leasing transformed premises on long-term bases). They may have purpose-built facilities on the hospital campus or they may be in stand-alone facilities elsewhere. Stand-alone facilities normally have special arrangements in place with the local hospital emergency response unit (crash team) so that in the unlikely event of an unforeseen medical emergency concerning a volunteer, help is guaranteed to arrive promptly. These commercial units are able to carry out the full range of Phase I studies required, including:

- First administration
- Rising dose tolerance
- Pharmacodynamics
- Pharmacokinetics
- Radiolabeled studies
- Bioavailability/bioequivalence
- Studies in special populations
    - elderly
    - renally, hepatically impaired
    - postmenopausal women
    - studies in patients not specifically related to treating the disease
- Drug/drug interaction studies
- Food/drug interaction studies
- Long-term tolerance studies

The larger units (with more than one hundred beds) function like well-oiled machines and run production-line-like studies on conventional drug products. Some of the more focused and specialized units have considerable experience with specific product types; indeed, many have administered drugs of certain classes much more frequently than any of their clients. These companies add real value to their clients' projects by virtue of this experience. Some units specialize in particular types of medicine (central nervous systems [CNS], for example), while others specialize in the recruitment of the special panels frequently required later in drug development programs, although they are still strictly part of Phase I studies.

Managing elderly volunteers can be particularly challenging to a Phase I unit. These volunteers, often eager to participate in studies, need significantly more care and attention than the standard young, healthy males who comprise the majority of early-stage volunteers. The typical youth hostel atmosphere is often unacceptable to the elderly. When choosing a CRO Phase I unit to carry out a study on elderly volunteers or on volunteers who are institutionalized for extended periods, particular attention must be paid to the social welfare conditions under which the study takes place. Hospital-based units are unlikely to be generally suitable.

The Phase I units in the United States and Canada appear to be of somewhat different character from those in Europe. First, they are substantially larger; no European unit exceeds 100 beds, whereas several American units (e.g., Phoenix, Harris, and Pharmaco) have over 250 beds. Second, the American companies seem to concentrate on the more routine studies (e.g., bioavailability, standard pharmacokinetics), while the European companies emphasize nonroutine studies (first time into man, and so on) and look to add value by assisting in drug development planning (*Technomark Register* 1999a; 1999b; 1999d).

Because cost is of crucial importance to generics manufacturers, some of them have historically had special relationships with Phase I CROs. Increasingly, these purchasers are looking to suppliers in nonmainstream countries such as Bulgaria, South Africa, and Ireland where suppliers typically charge less for their services than first-world-based suppliers.

PRECLINICAL AND TOXICOLOGY (PC&T) SUPPLIERS

Preclinical development covers a number of disciplines, including:

- Drug discovery
- Screening
- Pharmacology
  - primary and secondary animal pharmacology
  - absorption, distribution, metabolism, and excretion (ADME)
- Toxicology
  - acute toxicity
- Short-term animal studies (measured in weeks), including dermal and inhalation toxicology
- Longer-term/chronic toxicity
  - longer-term animal studies (six months to two years)
  - reprotoxicology, oncogenicity
- Genetic toxicology
  - in vitro toxicology/mutagenicity

"Regulatory toxicology" is a term frequently used to describe the body of tests required to gain permission to test the New Active Substance or New Medical Entity (NAS/NME) in volunteers and patients. Testing NAS/NMEs in humans is necessary to satisfy regulatory authorities in numerous countries to obtain approval to market the product. Most recently, the required studies have been laid down in guidelines set forth through the ICH (International Conference on Harmonization) process. The studies required are now more or less identical for all developed countries.

As described in the Appendix, evaluating the effects of NAS/NMEs in the numerous animal and cellular systems referred to previously requires a large volume of laboratory analytical work. The NAS/NME and its metabolites must be measured in different cellular, organ, and fluid media. Similar types of assays are developed and validated when measuring a drug and its metabolites, whether derived from animal or human samples. Some organizations are purely analytical companies, undertaking analysis of samples from all aspects of preclinical and clinical drug development. Other organizations undertake analysis of clinical samples only. Still other organizations analyze pathology samples from clinical trials on a centralized basis only.

In the preclinical area, the market is fragmented between those contract organizations providing services in discovery research, those companies providing in vitro and in vivo pharmacology and toxicology, and those companies providing analytical services. For example, companies referred to as "platform technology" companies in the biotechnology sector provide services in the discovery research area (genomics, proteomics, combinatorial chemistry, high throughput screening, computer-aided molecular design, and so on). Others provide in vitro and in vivo pharmacology and toxicology services with or without parallel analytical capabilities, while yet others have extensive analytical capabilities.

The PC&T market is generally mature and the majority of players are substantial companies. Historically, the market has been and continues to be concentrated in the United Kingdom (with five major facilities) and in North America (now with eight major facilities, two in Canada and six in the United States). Most other European countries have a single major facility; France has two. Japan also has a number of major facilities, but some of these are less than fully independent and virtually all of them supply the local Japanese market only.

Unlike clinical research, preclinical and toxicology studies do not have to be carried out in any specific geographical location. The United States and Canada undertake the lion's share of the PC&T work performed, though they are home to only 34 percent of the world's preclinical facilities. This disparity reflects the larger size and scope of these facilities within North America.

As an industry, the preclinical sector is much more mature than the clinical market. Some companies made strategic decisions to contract out large sections (up to 50 percent) of work to PC&T CROs over twenty years ago, resulting in the evolution of the key players in major pharmaceutical markets.

Probably because of its greater maturity, the PC&T market is growing more slowly than the clinical market today; we estimate the PC&T annual growth rate to be 8 percent. The proportion of preclinical R&D outsourced to CROs is a relatively constant 12 percent. This figure is likely to change only if CROs take over more pharmaceutical company preclinical development sites. Given the likelihood of continuing mergers between pharmaceutical companies in the future, this scenario is quite possible.

There is a dichotomy between the United States and Europe, in that the United States' pharmaceutical companies see their PC&T subcontracting as stable or decreasing (although biotechnology may take up some of the slack), while the European companies see their PC&T subcontracting to be stable or increasing over the next few years.

Table 1.5 shows the top twenty companies in the PC&T market as defined by 1998 revenues. Revenues for the publicly traded companies come from generally available information or, for privately held or nonprofit organizations, have been derived from internal assumptions on the basis of staffing levels. These companies represent 60 percent of the current market of $1.2 billion.

Table 1.5   **Leading Global Preclinical CROs 1998**

Rank	Company	Location of Facilities
1	Covance*	United Kingdom/United States/Germany
2	Battelle	United States
3	Huntingdon Life Sciences Group	United Kingdom/United States
4	PRIMEDICA (Genzyme Transgenics)	United States
5	Clintrials BioResearch Ltd*	Canada
6	Inveresk Research*	United Kingdom
7	Quintiles Preclinical Services*	United Kingdom
8	TherImmune Corporation (R.O.W.Labs)	United States
9	Phoenix*	France/United States
10	RCC Ltd	Switzerland
11	SRI International	United States
12	Oread Inc	United States
13	SafePharm Laboratories Ltd	United Kingdom
14	TNO-BIBRA	Netherlands/United Kingdom
15	Research Triangle Institute	United States
16	WIL Research Laboratories	United States
17	Ricerca Inc	United States
18	Centre International de Toxicologie	France
19	Biology & Zoology Research Center	Japan
20	Shin Nippon Biomedical Laboratories	Japan

*These companies have significant nonpreclinical activities in the clinical development area.

(*Technomark Register* 1999a; 1999b; 1999d. Reprinted with permission from Technomark Consulting Services Ltd.)

As stated previously, unlike the clinical market, the PC&T market is more mature and hence, the majority of players are substantial companies. In particular, the areas involving studies in live animals have high capital cost barriers to entry and smaller units can survive only in specialist in vitro testing areas. While a clinical CRO may typically have fifty to one hundred employees, with few exceptions, this is the minimum feasible size for a PC&T CRO.

Table 1.5 demonstrates that several of the larger companies are part of a global, full-service CRO. Traditionally, the preclinical area has been regarded as separate from the clinical area. Although some CROs, notably Quintiles, have made some effort to provide an apparently seamless transition from preclinical to Phase I study, these efforts have not been successful. Other multinational CROs see preclinical study as irrelevant to the rest of their offerings and have decided to divest themselves of their preclinical subsidiaries (e.g., PPD) or not to invest in the field (e.g., Parexel, Omnicare).

Few pharmaceutical companies themselves have drug development processes that move seamlessly from screening to Phase I studies. Decisions about which CRO will carry out preclinical work in general are made without regard to the considerations of the clinical studies to be (or that are being) carried out. Until the industry adopts a more holistic approach to drug development, the CROs that offer "full service" will find their message falling on deaf ears.

PHARMACEUTICAL DEVELOPMENT SUPPLIERS

The journey from synthesizing a few micromoles of an active substance to commercially producing a drug in ton quantities is a long and often tortuous one. Traditionally, development work was jealously guarded as an internal discipline exclusively. Companies that can turn synthetic routes in laboratory notebooks into robust processes for the production of kilogram quantities are relatively recent innovations, although there is now a growing number of small- to medium-sized chemical companies offering this service, both in the United States and in Europe.

Chemical synthesis development is clearly an area where suppliers can add real value to a chemical. The development of a new synthetic route with lower cost and greater efficiency can yield cost savings immediately translatable to a company's bottom line. This is an area where an innovative supplier can strike up a truly strategic relationship with great benefit to both parties.

Chemical synthesis is one field where the pharmaceutical industry shares some similarities with other industry sectors. For instance, pilot plant or full-scale production may well be carried out for one major company by its apparent competitor. Much in the same way that Ford may produce engine components for Renault, so Schering-Plough may synthesize Active Pharmaceutical Ingredients (APIs) for Merck or Glaxo-Wellcome. The commercial production scene is, however, beyond this book's scope.

Synthesis of the active substance is only part of the story. Generally, a more or less complex formulation must be developed, physically tested, and finalized before starting the pivotal Phase III studies. As with initial scale-up, the majority of this work traditionally has been carried out internally by companies. More recently, subcontractors or strategic partners have incorporated their innovative technologies into the drug development scene, not only adding value, but in some cases making a project viable. Historically, too many otherwise good drugs have failed because of the lack of a suitable formulation that would give the drug acceptable bioavailability or pharmacokinetics. Nowhere is this more true than in the protein field, where no proven oral delivery system exists yet for drugs such as insulin.

Formulation development suppliers thus include competitors, specialist companies with proprietary technology, and more conventional contract manufacturers. Suppliers of excipients can sometimes be a further source of assistance. These companies may provide technical service to pharmaceutical companies, on the assumption that when the product is commercialized, they will be the preferred supplier of the excipient. This service is often offered free of charge. The nearer a product is to commercialization, the more likely such a supplier is to provide the service; this is especially likely to be true for manufacturers of generics. Other formulation development suppliers include specialist companies, such as manufacturers of aerosol gases, metered dose inhalers, or filled syringes.

CLINICAL PATHOLOGY, CHEMICAL ANALYSIS, AND ASSAY SUPPLIERS

Chapter Ten, "Contracting Out Laboratory Analysis," deals with the selection and management of pathology laboratories. Many of the same considerations apply when choosing and managing a laboratory for the chemical analysis or assay of formulated product or of samples taken in Phase I or later clinical trials.

When clinical trials commenced on a significant scale in the mid-1960s, patient specimens were sent to the same laboratories that they would have been sent to if the patients were not in the trial. These laboratories varied by type, according to the country in which the trial was taking place. Thus in the UK, the majority of samples were sent to hospital laboratories. In France, samples went to: a hospital; the local *laboratoire d'analyses,* of which there were (and still are) several thousand; or, if esoteric analysis was required, to a laboratory in a network of official "quasi-state" reference laboratories. In Germany, there were fewer small, local laboratories and more regional laboratories that received samples from a large city or a *"land."* There were, of course, hospital laboratories as well. The bigger commercial laboratories were private, not state owned. The United States' system reflected the German paradigm, in that there were relatively few local laboratories, quite a number of laboratories that served a single city or state, and a growing number of laboratories that provided a national service. All the U.S. laboratories were privately owned.

As trials grew larger and electronic data handling was introduced, sponsors of clinical trials sought uniformity. Thus, the concept of a national central laboratory was born. This laboratory collected samples (actively or passively) from the entire country in which a trial was being conducted, analyzed them, and reported all the results to the doctor and the sponsor quickly and effectively.

In the United States, this trend prospered because of the preexisting national laboratory concept and because these companies were acquisitive, so that, by the 1990s in the United States, an oligarchy comprising a handful of big central laboratories had grown up. These included SmithKline Beecham Laboratories, LabCorp (originally a subsidiary of Roche), and Corning Laboratories Services. SciCor was the exception to this model. SciCor was set up in 1986 as a specialty laboratory for clinical trials only; it did no testing on patients not included in clinical trials. It was based in a city that was the hub of an air courier company, thus facilitating the transport of samples in and sample kits out.

As the market grew, Corning Inc. decided to acquire more laboratories for its new clinical trials laboratory service division. It also acquired a number of conventional CROs that offered clinical research, toxicology, and other services. In June 1991, Corning acquired SciCor, adding it to its CRO division. Corning then established a European laboratory by building a laboratory in Geneva. In June 1994, Corning acquired the Nichols Institute (a nationally known reference laboratory), which was not part of its laboratory services group but that did testing on behalf of SciCor whenever SciCor did not have the necessary expertise or experience in house.

Identifying revenues from companies offering central laboratory services (CLS) is difficult for the following reasons:

- The major players' business is part of a larger business—either a CRO or a general pathology laboratory.
- Many of the smaller companies are private and do not reveal their revenues.
- Where central laboratories are owned by CROs, the fees for that service may be incorporated in the fee for the complete clinical trial.

The major CLS players are:

- AAI/LAB
- BARC
- BioScientia
- ClinServe
- Covance
- CRL–Medinet Analytico
- Icon Laboratories
- InterLab
- LabCorp
- MDS/Cerba Glarif
- MR–CRL
- Nova Medical
- Q–Labs (Quintiles's brand name)
- Quest/SB

The total CLS market is estimated to be $600 million. The combined estimated revenues of the international companies above is $380 million. Companies operating in Europe alone are estimated to have revenues of $130 million; those operating in the United States and Canada generate estimated revenues of $70 million. Small laboratories in Australasia, South Africa, and Latin America may total $10–15 million.

The current CLS market is thus highly concentrated among a few large, intercontinental suppliers. Industry experts expect these companies to continue to dominate and, eventually, to wrest all the international business from national and regional suppliers. These local suppliers will continue to win business from companies running studies for regional purposes, for local Phase IV studies, and for smaller specialist Phase II studies. As esoteric testing increases, the local laboratories will struggle even to get the local studies, unless they have a large standard medical business.

Suppliers of analytical chemistry fall into five categories:

- Phase I companies;
- Toxicology companies;
- Pathology laboratories (hospital and specialist);
- Divisions of pure analytical companies that offer services to a variety of industries; and
- Global multiservice CROs.

Analysis is required throughout the drug development phase, including:

- During synthesis and scale-up;
- In Phase I studies, as part of pharmacokinetic (PK) and human ADME studies;

- In animal toxicology and pharmacology studies; and
- In formulation development.

Analysis may also be required when active substances or formulations are imported into a particular country.

These five categories of analysis are performed by a wealth of suppliers. Today Liquid Chromatography–Mass Spectrometry–Mass Spectrometry (LC-MS-MS) is becoming an increasingly used analytical technique for all the purposes in the preceding list, in particular for low concentrations measured in PK and ADME studies. Such equipment requires substantial capital investment and, consequently, smaller suppliers will find it increasingly difficult to compete as the technique becomes an industry standard.

POSTREGISTRATION (PR) STUDIES AND PHASE IV (PR STUDIES) SUPPLIERS

Phase IV Studies

Defined simply, Phase IV studies are those performed on a drug postregistration. However, whereas trials exploring new indications with a registered drug, new methods of administration, or new combinations were all considered Phase IV studies in the past, the new EU GCP guidelines state that Phase IV trials are carried out on the basis of information in the marketing authorization summary of product characteristics. On this basis, the other studies mentioned above now are considered trials for new medicinal products.

In today's climate, therefore, a Phase IV trial may be defined as interventional and carried out using a licensed formulation within the terms of its product license. Furthermore, it may be conducted either in general practice or in a hospital, primarily to extend the efficacy database, although collection of safety data forms an essential part of such a study as well. Most trials employ an active comparator, and the sponsor supplies all clinical trial materials. A Phase IV trial therefore differs from a postmarketing surveillance study, which is observational, noninterventional, and conducted primarily to monitor safety when a newly introduced medicine is prescribed in everyday clinical practice (see below).

In the past, uncontrolled marketing studies, planned largely for marketing advantage rather than detection of new adverse drug reactions (ADRs) or demonstration of further efficacy, were also considered Phase IV studies and described as seeding trials. The underlying aim of seeding trials was to influence physician prescribing, although this factor went unmentioned in the protocol. Seeding trials were conducted without the benefit of control groups, adequate statistical power, or meaningful safety and efficacy design. Happily, these quick and dirty studies are a thing of the past, discontinued on scientific, ethical, and economic grounds. In addition, important changes in the healthcare environment (health economics and quality of life measurements) have led to a shifting in the key roles of healthcare decision makers.

Postmarketing Surveillance

The key objective of postmarketing surveillance studies is evaluating the safety of newly licensed drugs, although simple efficacy evaluation measures and comparator drug observation may also be incorporated.

However, to establish the nature of these studies more fully, it is necessary to take a step back to the historical objective of postmarketing surveillance activities. The term "postmarketing surveillance" itself may in fact mean different things to different people, depending on their point of view. Postmarketing surveillance can be confused with tracking the spontaneous, ongoing reporting of ADRs once a product has received a marketing authorization. This kind of ADR tracking is sometimes described as pharmacovigilance. The expected frequency of such reports may depend not only on the nature of the product and the licensing authority (some products, such as biotech or particularly novel products, are subject to more stringent monitoring requirements imposed by regulatory authorities when they are first marketed), but also on the extent of underreporting (due to lethargy, ignorance, fear of litigation, and so on). Spontaneous ADRs originate from three sources primarily: directly from the notifying physician, from the sponsor pharmaceutical company, and from reports arising in the literature (including the lay press). In addition, ADRs arising during Phase III clinical trials (for example, of a marketed product in a new indication) or during the course of structured postmarketing surveillance studies are also considered spontaneous reports.

Bona fide postmarketing surveillance studies have been known to take several forms, with potentially several drawbacks, depending on the drug in question and the circumstances at the time.

Cohort studies, on the other hand, identify and observe patients receiving a drug over a period of time to determine the rate and outcomes of ADRs. Another group of patients not receiving the drug, or possibly receiving another drug, are identified for comparison. Cohort studies can be performed prospectively (as the study is ongoing) or retrospectively (once the ADR has occurred), using medical records, questionnaires, and interviews. These studies resemble controlled clinical trials and can be used to generate new ADR hypotheses; however, in cohort studies, unlike controlled trials for patients, the decision about treatment is made prior to a patient's inclusion, an activity that may involve a channeling bias (e.g., the patient suffers worse symptoms than the average patient with the disease or has been unresponsive to previous treatment). In addition, the selection of an inappropriate control group can easily influence the interpretation of results.

Health Economics and Quality of Life Measurements

Health economics, writes Kobelt (1996), is a "discipline defined as the application of the theories, tools and concepts of economics to the topics of health and healthcare. Since economics as a science is concerned with the allocation of scarce resources, health economics is concerned with the allocation of scarce resources to improve health. This includes both resource allocation within the economy to the healthcare system and within the healthcare system to different activities and individuals."

Health economics and pharmacoeconomics are sometimes used interchangeably, although some observers would use the term "pharmacoeconomics" to describe strategies focused on the quality of life and overall benefit to the patient for a compound once it's registered. In this sense, pharmacoeconomics is seen as referring to drugs, whereas health economics is seen as referring to healthcare more generally. These understandings contrast with perhaps a more widely understood view of health economics as a pure preapproval accounting exercise (cost of treatment *x* versus cost of treatment *y*) to position a product within a formulary/price range.

Health economics is concerned with the business of measuring cost-effectiveness—increasingly a primary concern to the pharmaceutical industry. During the last two decades, there have been many examples of pharmaceutical cost-effectiveness reducing healthcare-sector costs or, instead, demonstrating that a more expensive medicine may prove more cost-effective than a less expensive one.

Three important interrelated changes have created interest in quality of life assessment. First, chronic disease (with the exception of AIDS) is rising in prominence as a primary target of medical intervention because healthcare is increasingly viewed not as a vehicle to save lives but, rather, as an opportunity to prolong life or improve the quality of the remaining survival time (i.e., adding life to years rather than years to life). The second factor is the shifting roles of key healthcare decision makers, with responsibility shifting downward from the doctor to the patient and upward from the doctor to emerging healthcare managers. Both doctors and managers are interested in the net benefit of the medicine and are concerned that new therapeutic regimens should be seen as cost-effective before the drugs are introduced and reimbursed to the patient. The third and final factor is a dramatic transition in the medical practice ground rules—the demand that medical practice give "value for money."

Quality of life, with regard to measuring the benefits of medicines, is defined as the level of well-being and satisfaction associated with an individual's life and how these are affected by disease, accidents, and treatment—so-called health-related quality of life. A patient's quality of life can be measured by two broad methods: the health profile system and the health index. The health profile system measures different parameters for individual patients or diseases (e.g., sleep, emotion, pain, physical mobility), while the health index attempts to measure patients' well-being on a single numerical scale from a consensus of people's self-evaluations.

A controversial topic allied to health economic and quality of life assessments is a concept called evidence-based medicine—the attempt to ensure the translation of evidence of a new treatment's efficacy and cost-effectiveness directly into changed clinical behavior and better patient services. Clinicians and economists alike vigorously debate the potential benefits of considering the individual patient or a patient population.

Suppliers of PR Studies

Typically, the pharmaceutical companies conduct PR studies themselves. However, they are beginning to use external suppliers more frequently, due in part to the perceived impartiality of such suppliers. PR suppliers include:

- Global CROs (often specialist divisions);
- Local CROs;

- Specialist PR-studies companies;
- Government/university institutions; and
- Market research organizations.

Estimating the total market size for PR studies has proven difficult, partially because global companies do not break down their revenues appropriately. Very approximately, we estimate that all PR studies represent less than 10 percent ($350 million) of the Phase II/III market, with the majority of the market being in the United States.

Methodologies are evolving rapidly in this market. The need for economics-based drug assessment has never been greater, given the financial pressures under which all healthcare provision systems struggle. Unfortunately, there still appear to be many companies that dabble in PR studies. *Technomark Register* (1999a; 1999b; 1999c; 1999d) lists over 200 in Europe and 220 in North America. The wise purchaser will select from those few good-quality specialist suppliers that remain independent, from the specialist divisions of the multinational CROs, or from selected university/government institutes.

Here we can only scratch the surface of this topic.

REFERENCES

Hughes, R. G. and C. E. Lumley. 1999. *Current strategies and future prospects in pharmaceutical outsourcing.* London: Technomark Consulting Services Ltd.

Kobelt, G. 1996. *Health economics: An introduction to economic evaluation.* London: Office of Health Economics.

Technomark Register: European Contract Research Organisations—Clinical Research. 1999a. Vol. 1. London: Technomark Consulting Services Ltd.

Technomark Register: European Toxicology and Analytical Organisations. 1999b. Vol. 2. London: Technomark Consulting Services Ltd.

Technomark Register: Contract Packers and Manufacturers—Europe. 1999c. Vol. 3. London: Technomark Consulting Services Ltd.

Technomark Register: Contract Research Organisations—North America. 1999d. Vol. 4. London: Technomark Consulting Services Ltd.

Roy Drucker

The Sourcing Decision

The one who adapts his policy to the times prospers, and likewise that the one whose policy clashes with the demands of the times does not.

Machiavelli, *The Prince*

There is no art that has been more cankered, more sullied with aphorising pedantry than the art of policy.

Milton

It is no accident that this chapter begins with two quotations illustrating the necessity for a company to have a policy relevant to the times and pointing out that, for centuries, the notion of policy has been much misunderstood. A company must perforce have a sourcing policy that stems from the repeated application of a set of actions derived from a strategic position. Indeed, so far as a company is concerned, a rational sourcing decision in the absence of a sourcing policy is an oxymoron!

The purpose of this chapter is to delineate some key dimensions that should be encompassed in those deliberations leading to a sourcing policy. As has been indicated, a specific sourcing decision should flow quite naturally from the policy framework.

Sourcing policy in pharmaceutical research and development (R&D) and the associated decisions have become more prominent with the growing realization that nothing less than the entire effector arm for the company's future products is at stake. Gone are the days when a company's working assumption was that "our internal capabilities are always superior and we will only go outside when we are swamped."

In achieving R&D objectives, the avenues available to a pharmaceutical company are threefold:

- Do everything internally with existing internal personnel—resource;
- Do everything internally with extant internal personnel plus import staff on a temporary basis as needed—insource; or
- Place work with an external organization—outsource.

Historically, the major commitment has been to resourcing and this is still true today, but outsourcing is growing inexorably and, *pari passu* with this, an ample supplier base. Worldwide sales of contract clinical research in 2002 will exceed $5 billion; the preclinical contract market is about half this size.

SUPPLIER ANALYSIS BY SECTOR

The sectoral breakdown of the CRO market by annual revenue is shown in Figure 2.1.

Table 2.1 shows approximate breakdowns of the various stages of drug development on a worldwide basis. These figures must, however, be interpreted with caution; most CROs do not distinguish in their revenues between these various sectors. There is, therefore, some arbitrary assignment of these revenues to the sectors.

Estimating shares of this market is particularly difficult since much of the revenue, particularly in Europe, derives from small parts of very large private companies (clinical pathology companies).

Technomark has estimated, from a variety of industry sources, the CRO growth rates by sector. These estimates are shown in Figure 2.2.

Figure 2.1   Breakdown of global CRO market

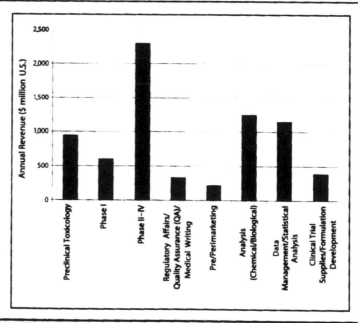

(Compiled from Technomark internal data. Printed with permission from Technomark Consulting Services Ltd.)

Table 2.1   Percent of CRO Market Share in Drug Development Stages by Territory

Territory	Phase I (%)	Phases II–IV (%)	Data Management and Statistics (%)	Regulatory Affairs (%)	Postmarketing (%)
U.S.	37.5	63	65	44.0	84
UK	21.5	12	14	15.5	6
France	7.5	6	4	14.5	5
Germany	12.5	7	8	12.0	4
Rest of Europe	11.5	8	8	11.0	1
Rest of World	9.5	4	1	3.0	0
Total Market $ million U.S.	600	2,300	1,150	330	220

Territory	Analytical (%)	Clinical Trials Supply (%)	Regulatory Pharmacology & Toxicology (RP&T) (%)
U.S.	63	55	57
UK	13	22	28
France	4	5	3
Germany	8	5	4
Rest of World	12	13	8
Total Market $ million U.S.	1,250	360	950

(Compiled from Technomark internal data. Printed with permission from Technomark Consulting Services Ltd.)

Figure 2.2   Estimated annual growth rate of market sectors

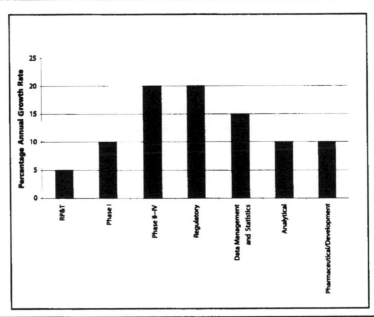

(Compiled from Technomark internal data. Printed with permission from Technomark Consulting Services Ltd.)

For 2000, a weighted average of these growth rates gives a mean growth rate of 15 percent. This is lower than the historical growth rate, but it is the general feeling that this will return to a figure of 18–20 percent if and when the current phase of pharmaceutical industry consolidation activity settles. There is no particular difference in growth rates between the United States/Canada and Europe. Europe has, in general, outsourced its development rather more reluctantly and more recently than the United States. However, since many of the top twenty pharmaceutical companies have development centers on both sides of the Atlantic, European acceleration of outsourcing into the United States seems to be compensating for any slowing down in the United States. Clearly, as pharmaceutical companies reach what they regard as an optimal level for outsourcing of their work, the growth rates of the CROs will slow down. Moreover, there is the slowing down of what we have described as a leverage effect. Thus, if CRO revenue of $7 billion represents currently some 20 percent of overall R&D, an increase of this figure to 25 percent of an R&D budget that is increasing at 5 percent yearly will translate into a market growth of the CRO of 31 percent. To illustrate, in Table 2.2, figures are projected for the current and next few years, showing the decreasing influence of this leveraging effect.

Technomark's predicted growth rates do not take into account any potential major strategic shift in the outsourcing of R&D from one of the top twenty R&D spenders. If one of these were to go from a ratio of 80/20 to 20/80, this would accelerate growth rates at a stroke.

The most important factors that have influenced drug companies to choose to contract out work and that have created the recent rapid expansion of the CRO industry are:

- Regulatory and legislative changes requiring pharmaceutical companies to perform larger, more complex, and more wide-ranging studies;
- The introduction of Good Clinical Practice (GCP), which has sharply increased the work required for a regulatory submission (historically very important, but now much less significant);
- Globalization of the pharmaceutical industry (local regulatory, linguistic, and cultural knowledge makes the use of CROs very attractive);
- The growth of the biotechnology industry, which has little experience of the complex process of the clinical development of a drug;

Table 2.2 Projected Leveraging Effect

Year	R&D* Budget ($ billion U.S.)	% Outsourced	CRO Market ($ billion U.S.)	Yearly CRO Market Growth Rate
2000	35.00	20	7.00	—
2001	36.75	25	9.19	31.3
2002	38.59	30	11.58	26.0
2003	40.52	35	14.18	22.5

* = 5 percent growth.

(Table generated by R. G. Hughes. Printed with permission from Technomark Consulting Services Ltd.)

- "Strategic" decisions by many pharmaceutical companies to use CROs for certain areas of R&D to try to reduce costs or to enable minimal staffing in overseas offices;
- With wider acceptance, an increasing value put on CRO services by the pharmaceutical industry;
- Increasing activity by Japanese companies overseas; and
- Pharmaceutical industry consolidation.

These issues are more comprehensively reviewed in the following chapter.

BENCHMARKING AND CORE COMPETENCES

An explicit sourcing policy will recognize how the sourcing options are to be incorporated into overall sourcing, if at all. It is important to acknowledge that there is no one-size-fits-all solution. There is a compelling case for benchmarking exercises to identify best practice in sourcing within the industry and, probably more profitably, beyond it, but such activities must then be reconciled with the realities of a particular company, its history, tradition, cultures, current situation, and future aspirations.

It is helpful if an organization can think of those who undertake activities as suppliers, be they internal or external. A best practice sourcing policy will identify the different types of relationships with suppliers, generally conditioned on the risks related to the activity performed and its commercial significance (i.e., its ability to impact value) and the structure of the supply sector (e.g., many or few specialist suppliers). It will also be based on clearly defined core competences.

With respect to core competences and sourcing policy, the issue is not what core competences a company has, but those it will need to support an optimal sourcing policy. Thus, core competence identification necessitates a level of forward planning so that future internal resources will sustain the sourcing policy that is approved and implemented.

Core competences will be different in mix for each company. They may relate to management skills and geographies, as well as scientific or medical areas. They will evolve over time. It is critical that future core competences are agreed upon at senior management level and that the whole organization works toward realizing them in the required time frame. In some cases, selection will be by omission and senior managers will agree upon what are not core competences. The number of core competences needs to be limited to those in which the company can continue to invest to maintain a leadership position. Core competences will generally be defined in terms of:

- The factors that differentiate the company from its competitors;
- The activities without which the company would be at a considerable competitive disadvantage;
- The factors that are the heart and drivers of the business; and
- The factors that drive the value of the business.

Quite specifically, to qualify as a core competence, it is not enough that a large volume of an activity is undertaken, that a large amount of money is spent on the activity, or even that it has been undertaken for a long period of time or that it coincides with the disciplinary background of one or more senior management advocates.

The matter of core competence has been labored because, without the requisite core competences in place, the best sourcing policy will be associated with suboptimal implementation of sourcing decisions, if not worse!

DRIVERS TO THE SOURCING CHOICE

De facto, for a given activity, a sourcing decision is required, preferably by commission, but otherwise by default. The decision tree is typically as displayed diagrammatically in Figure 2.3.

The drivers for undertaking activity externally and outsourcing work, typically to CROs, have been categorized by Greaver (1998) as follows:

Drivers for Outsourcing

Organizational drivers

- Enhance effectiveness;
- Increase flexibility;
- Transform the organization; and
- Increase value.

Improvement drivers

- Improve operational performance;
- Access external skills;
- Improve management control;
- Acquire ideas; and
- Improve credibility.

Financial drivers

- Reduce investment; and
- Generate cash.

Figure 2.3   Sourcing decision tree

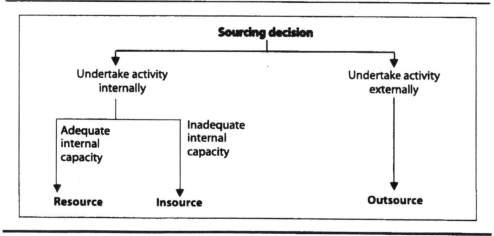

Revenue drivers

- Gain market access via provider;
- Accelerate expansion;
- Expand sales and production capacity; and
- Exploit existing skills.

Cost drivers

- Reduce costs via providers' specialization; and
- Turn fixed costs into assets.

Employee drivers

- Stronger career paths; and
- Increase commitment.

These drivers apply broadly to pharmaceutical R&D and can be drivers to either strategic or tactical outsourcing. How they actually impact on the company's outsourcing will in fact be dependent on the overall sourcing strategy the company decides to implement.

Surprisingly, it is challenging to list positive drivers for undertaking, as opposed to directing, activity internally. Such lists can easily become an indictment of management's ability to manage external relationships. Many of the perceived benefits of internal activity are diminished when true strategic outsourcing, rather than the far more commonplace tactical outsourcing, is available as an alternative.

The basic characteristics of tactical outsourcing do not differ significantly from typical buyer-seller relationships pertaining to any service or product. Tactical outsourcing takes place currently for a number of reasons. The drivers may be either due to lack of internal capacity or the availability of external opportunity (see Table 2.3). It is when the key drivers lie more at the external opportunity end of the spectrum that outsourcing is likely to become less tactical and more strategic.

Drivers for Tactical Outsourcing

Typical examples of internal drivers are:

- Internal shortage in
  - clinical monitoring capacity; and/or
  - clinical trials supply manufacturing capability.
- Lack of experience
  - therapeutic experience with a particular disease state;

Table 2.3 Factors Leading to Tactical Outsourcing

Internal Shortage	External Opportunity
• capacity	• expertise
• experience	• experience
• expertise	• equipment
• equipment	• risk spreading
• time	
• money	

- ◆ analytical experience with the type of therapeutic agent; and/or
- ◆ experience running trials in a particular country
- Lack of expertise
  - ◆ in a required analytical technique; and/or
  - ◆ in formulating a novel compound as a dosage form.
- Lack of equipment
  - ◆ access internally to major analytical or diagnostic equipment (e.g., MRI, PET, NMR, HPLC, AMS); and/or
  - ◆ access to synthetic chemistry methodology or equipment (e.g., fluorination/high pressure hydrogenation).

Whereas the drivers for tactical outsourcing are essentially internally derived, the drivers of strategic outsourcing are some of the major issues confronting pharmaceutical companies in the Twenty-first Century.

Major issues for pharmaceutical companies in the Twenty-first Century include:

- Critical mass of company*;
- Direction of research;
- Advances in genetics and genomics*;
- Novel platform technology emergence*;
- Changing medical needs*;
- Increasing numbers of drug candidates*;
- Building value in the development process*;
- Healthcare spending reduction and redistribution*;
- Patient group pressure and consumerism;
- Evidence-based medicine;
- Pharmacoeconomics and "fourth hurdles";
- Increasing cost and time to market*;
- Decreasing patent exclusivity*;
- Regulatory harmonization*;
- Self-medication and alternative medicine;
- Healthcare restructuring;
- Generic competition*;
- Changing population demographics*;
- Information technology and the World Wide Web*;
  * = Drivers of strategic outsourcing

Tactical outsourcing is generally understood to be effected at short notice on a project-by-project basis. It is often required when internal resources of manpower or capacity are inadequate. It may be opportunistic, in that external onetime opportunities to obtain a service or a product that would otherwise be produced internally may be grasped eagerly if the price or delivery conditions are advantageous. In a pharmaceutical development context, a CRO may be used on a tactical basis, for instance, if it has had a clinical trial cancelled by a different sponsor, having lined up investigators to run the trial. Use of the CRO can thus save money and time for the sponsor compared with having to recruit investigators afresh for a trial in the same or similar indication.

Tactical outsourcing is usually highly controlled by contracts (binding on both sides) and, hence, tends to be inflexible. Contracts are often very precise as to not only the deliverables, but also the methodology of their generation (be they a service or a product deliverable). Regrettably, tactical relationships are often confrontational rather than collaborative. Their short-term nature (in respect to the setting up of the contract, rather than its length) with no assurance of a long-term relationship tends to instill an "us and them" mentality—especially when things go wrong!

Because of the detailed contractual relationships, tactical outsourcing is driven by the price of the service or product. The cost to the sponsor, with respect to its internal cost plus the external price, tends to be a secondary consideration. Value in terms of how the supplier can deliver added value, especially in the case of a service, tends not to be considered at all. Thus tactical relationships lead to a commodification of the service or product.

Strategic outsourcing has an entirely different face. It is based on the evaluation of the long-term core competences of the sponsor rather than the short-term needs. It looks very much at the value of the product or service rather than, as we have just seen, the price or cost. This value focus is fortified by the long-term relationships between sponsor and supplier. These relationships foster trust between the parties, so that the deliverable is the important consideration and the mechanism by which it is generated is of much lesser importance.

It is a characteristic of strategic outsourcing relationships that both parties tend to treat each other as equals. This by no means implies that the parties are of equal size. Indeed, some of the better relationships that have developed in other industries are between corporate giants and small contractors. Big industry has understood that its corporate structure cannot engender the flexibility and adaptability that pertain to their smaller partners and so holds on to these partners almost jealously. Likewise, the small company enjoys the continuity of business that the corporate giant can provide. The relationship can, of course, be the other way around. Small entrepreneurial companies, unwilling to establish significant fixed overheads, may contract in a wealth of services from much larger suppliers. These suppliers are savvy enough to recognize that from these small entrepreneurial companies new, high-technology giants can emerge and, if they have developed truly strategic relationships, then these will survive the corporate growth and provide major business opportunities in the future for the supplier.

ORGANIZATION FOR THE SOURCING DECISION

While the organization charged with sourcing decisions will vary from company to company, there are certain commonalities to effective company structures:

- They recognize the responsibility for the sourcing decision to be separate from that for the subsequent management of relationships and general product development;
- Sourcing decisions are not the exclusive prerogative of functional managers but rather of a cross-functional team. The multifunctional team has responsibility for the specific sourcing decision;

- The team responsible for the sourcing decision has a clear reporting line that is more than the aggregate of the individual team members' line reports; and
- The team has a clear charge and clear boundaries.

Experience has shown that effective teams benefit from an absence of direction within the area of their charge. They will focus on the process whereby they arrive at sourcing decisions, as well as the content of them.

EVOLVING PHARMACEUTICAL SOURCING TEAMS THAT WORK WELL

Project management and project teams are well-established in the pharmaceutical industry. However, project management has historically been subordinate to functional management, and project teams have been known to function as committees with representation from different functions.

There is a need to devote resources to developing cohesive project teams with members identifying with the team rather than the function and rewarding the members on the basis of team performance, not to be confused with regulatory and marketing success. It is essential that there be metrics of team performance and that individual members receive feedback on their performance from other members of the team.

CRITICAL EVALUATION OF THE PHARMACEUTICAL PURCHASING FUNCTION

In a pharmaceutical company, purchasing functions traditionally had no responsibility for CROs/partnerships. As it became clear that large sums of money were being spent out-of-house on work assigned to CROs, "purchasing" claimed increasing responsibility, but often equated the CRO services with those of caterers, security companies, and so on. The message from other industries is that strategic purchasing and commodity purchasing cannot be equated and may not coexist well in the same department, certainly not at the same responsibility level. Pharmaceutical companies that have lumped together their purchasing of external functions need to examine the role of purchasing in relationship to the teams advocated above and, if necessary, create or elevate the status of these teams to reflect the critical nature of their function to the success of product development.

CREATE A COHESIVE, COHERENT, DECISION-MAKING STRUCTURE

Many pharmaceutical companies have processes and practices for the selection of CROs and other key suppliers. These will usually be enshrined in SOPs and implemented at various organizational levels. Companies need to reexamine their processes, with respect to strategic outsourcing in particular, so that not only are the mechanical aspects of the CRO assessed and decided upon but also that the added value attributes are appreciated and taken into account. In particular, price, where it is an issue, should be taken on board as part of a process and not as an item to be negotiated separately. It is important that these processes apply in the different parts of a company, so that suppliers perceive a coherent policy and can ally themselves uniformly to the company as a whole and not differently to different parts of it. Clearly, the level at which decisions on outsourcing partners are made should depend on issues such as the project's criticality as well as its overall price level—but the process should be codified to reflect this.

NATURE OF THE DECISION AND SUPPLIER ISSUES

The fundamental sourcing decision is a "make versus buy" one and will take into account the following:

- Availability of internal and external suppliers;
- Extent to which internal or external suppliers could be developed and made available in the required time frame;
- Technology and other resource constraints; and
- Risks and rewards associated with the activity.

There should be a clear set of criteria that incorporates the following major areas of consideration:

- Need for sponsor/supplier alignment, where necessary;
- Need for sponsor/supplier cultural fit, where applicable;
- Processes and practices;
- Core competencies of the pharmaceutical company; and
- Baseline quality, cost, and delivery expectations with different sourcing options.

DECISION AND SELECTION

The sourcing decision should be based upon the relative value creation possible. This, in turn, requires an understanding of the value chain, of which the activity to be sourced is a part. Pharmaceutical companies should be wary of value chains that substitute an external activity for an internal one and consider only relative costs—the uncritical commodification of drug development activities.

Specific outsourcing decisions should be taken or recommended by those with the technical competence to analyze the relative merits of available options. Pharmaceutical management may disempower its more junior staff by inappropriately referring tactical decisions to top management. However, the top-down decision-making structure must be in place so that specific programmatic decision making is taken in the context of the overall portfolio, while the program is the context for decisions relating to individual studies.

CREATE RESONANT VALUE CHAINS REFLECTING INTERNAL VERSUS OUTSOURCED OPTIONS

The recognition of value chains and clarity about where and how a company generates value is important. Equally, it is important for pharmaceutical companies and CROs to be aware of the places in the value chain where they are not creating value. Value creation emphasis should be not only on the basis of individual drug development activities, but also on synergies from the orchestration of activities to exploit their relatedness (i.e., the focus should be on the coupling). There is a need to decommodify the actual units of activity to be coupled.

There is a current vogue for activity-based cost accounting (ABC accounting) that unitizes activities and causes activities with a given designation to be equated without giving recognition to the inherent value that may be added, depending upon who undertakes an activity. There must be critical appraisal of the value added by different suppliers of an activity. It is insufficient to think merely in terms of price and quality.

In selecting a supplier, there are five key attributes that assume greater significance as a more strategic relationship is sought between sponsor and supplier:

- "Fit" with outsourcer—cultural, strategic, and organizational;
- Flexibility;
- Technology leadership and capability;
- Commitment to the overall success of the end product; and
- Alignment with customer focus.

CRITICAL APPRAISAL OF CONTRACTUAL BASIS

Typically, CRO contracts are both voluminous and rigorous, circumscribing the activities of the CRO and thus often inhibiting its ability to create value. In other industries, where strategic relationships are more common, relationships have evolved so that the contractual basis underpinning the work assigned to strategic suppliers is looser, although benchmarks, milestones, and design specifications may be tight. As trust develops (and typical pharmaceutical contracts enshrine mistrust), the pharmaceutical industry needs to examine its contracts to allow, where appropriate, for value to be added flexibly and cooperatively.

CONCLUSION

The sourcing decision is an inherently difficult one, even in the presence of an adequate policy framework. Nonetheless, consistent, coherent sourcing behavior is a source of competitive advantage. This is particularly the case when strategic outsourcing is leveraged to the full extent possible. There is a recognizable paradigm shift ongoing in the pharmaceutical industry, as more and more companies recognize the value of focusing on their core competences and working collaboratively with others to achieve their corporate goals in what has been termed the extended enterprise. This is reflected in the changing balance of sourcing decisions and will be dealt with in Chapter Fifteen, "Trends in Pharmaceutical Development Outsourcing," by Drs Rudolph and Tempio.

REFERENCE

Greaver, M. 1998. *Strategic outsourcing: A structured approach to outsourcing decisions and initiation.* New York: AMACOM.

Tim Wright

Supplier Identification, Evaluation, and Selection

Choosing the best suppliers requires a clear resourcing strategy—this has been demonstrated in Chapters Two and Four. Organizations equipped to focus their in-house resources on clearly defined and understood core activities *and* to utilize suppliers to greatest effect will have considerably enhanced competitive advantage in the long run (Quinn and Hilmer 1994). In spite of this observation, there is often significant resistance from organizations to outsourcing any activities. Hence, managers themselves must understand and be able to demonstrate the broad business advantages of an integrated resourcing strategy if outsourcing is to be accepted and implemented effectively.

Implementing a successful resourcing strategy itself requires commitment of human and financial resources. These should be targeted at understanding the core competencies of the company, the economics of delivering key outcomes, and how these are expected to change in future. This understanding helps identify which competencies can be best developed and retained by the company itself and which are best outsourced. The availability of expertise within contract research organizations (CROs) should be a key factor in determining those competences that should be maintained and developed in house.

The spectrum of activities identified as core will differ significantly between a small, focused company and a major pharmaceutical company. Identifying core competences is more complex in a large company because there are multiple departments and functions experiencing resource shortfalls at different times. Projects should be evaluated early in their life cycles by managers with appropriate seniority and experience, and coordinated across departments to optimize the use of in-house and external resources.

ESTABLISHING A FRAMEWORK FOR IDENTIFYING, SELECTING, AND EVALUATING CROs

With a clear resourcing strategy established, a framework for identifying CROs to meet short- and long-term needs must be established. Typically, this framework develops in three stages.

During the first stage, companies improve their rationale and processes for evaluating and selecting CROs for individual projects.

As the company develops, it may need to outsource more extensively and frequently. This leads to the second stage, in which a company develops a supplier base to meet the majority of its current and planned outsourcing needs. This process should be forward looking and sustained. This second stage allows a more relational approach to choosing suppliers, based on accumulated experience.

As experience is gained, the third stage should be the adoption of preferred partners who provide benefits as a result of close and frequent collaboration. The frequent use of such partners does not mean being tied to a limited supplier list but, instead, reflects successful collaborative relationships over time. This chapter expands upon the many important ways in which these three stages differ.

STAGE ONE: IDENTIFYING, EVALUATING, AND SELECTING CROs FOR INDIVIDUAL PROJECTS

Figure 3.1 summarizes the key steps to outsourcing individual projects or activities. It is assumed that the project is outsourced within an established resourcing strategy and that a cash budget is dedicated to pay for the project (see Chapters Two and Four).

Planning

Depending on the size and complexity of the project to be outsourced and on the maturity of the relationship with potential suppliers, sufficient time must be allowed to:

Figure 3.1   The key stages for outsourcing individual projects

1. Planning	2. Gather Data on CROs	3. Initial Evaluation of CROs	4. Request Proposals	5. Select CRO
Determine timetable	CRO capabilities	Basic capabilities	Standardize process	Select best CRO
Agree upon outsourcing team	Utilize best data sources	Project-specific capabilities	Evaluate proposals	Audit CRO
Define CRO responsibilities		Shortlist CROs	Resolve queries	Commence project
			Meet 2–3 best CROs	

- Define the scope of the project to be outsourced;
- Define exactly what the CRO will be responsible for delivering;
- Identify potential CROs;
- Engage suppliers in dialogue and request and evaluate operational feasibility reports and cost quotations;
- Meet with CROs to discuss and clarify feasibility reports and costs;
- Select the best CRO, negotiate, and agree on terms and conditions;
- For new CROs, undertake an audit of facilities, practices, and procedures;
- Sign a contract; and
- Initiate the project.

An experienced CRO can often achieve a faster project initiation than the company that is outsourcing. However, a realistic lead time must allow CROs to engage third parties, such as investigators, ethics committees, and regulatory authorities, prior to patients actually being entered into the study. This phase should be carefully managed since good planning and setup are likely to accrue efficiencies later.

Project initiation can be expedited by entering into a limited-scope contract associated with early project activities (e.g., contacting investigator sites, obtaining ethics committee and regulatory approvals). Such a contract can be used to bridge the time gap often associated with finalizing details of the full contract for a major project.

The outsourcing team

The relative importance of outsourcing will determine the type of team or organization assembled to support outsourcing. Companies outsourcing infrequently and sporadically may nominate a team comprising representatives such as the project leader or planner, along with finance or procurement department support. In contrast, companies that have a significant commitment to outsourcing may develop a permanent team or department to support all outsourcing activities. This should comprise staff with considerable operational experience who have worked with CROs before and understand the key aspects of procurement and finance. Such a group can provide critical support to the multidisciplinary team gathered to outsource a major project or function (see Figure 3.2).

For a project involving a number of functions (for example, clinical, data management, statistics, and regulatory activities), each over a number of countries, it will be necessary to consult and establish liaisons with the corresponding managers. These key departments should participate actively in the CRO evaluation and selection process. Responsibility and ownership of the project should not be allotted solely to the outsourcing manager or department, but, instead, should be accepted by the project team and its manager. This is important because, after signature of the contract, the CRO will deliver outcomes to the project team and not the outsourcing manager.

Not all team members are needed for all parts of the outsourcing and project management process. For example, performing the technical evaluation of a CRO requires a different skill mix than the mix needed to negotiate the financial terms of the contract.

Figure 3.2   Project outsourcing is managed by a core team who calls on the expertise of other departments

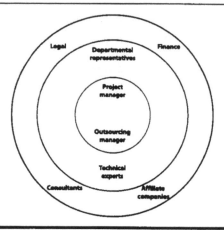

Defining responsibilities for the project

Before approaching any CROs, it is critically important to define, in writing, exactly what are the main CRO responsibilities. This will be achieved through consultation and consensus with the project staff, who are responsible internally for delivery of the various project components. The CRO and sponsor responsibilities and deliverables should be complementary and, together, define all deliverables essential for the success of the project.

Probably the simplest method to document this requirement is to develop a pro forma table listing all tasks necessary to complete a typical clinical, data management, statistics, or regulatory project. Responsibility and delivery time for each task are then assigned (see Table 3.1). The pro forma table can be used later to help define the scope and cost of the project (see "Requesting a Project-Specific Proposal from Shortlisted CROs" later in this chapter). Where appropriate, it is helpful to identify activities to be performed by the main sponsor company, its affiliate companies in specific countries, and the CRO.

With this overview in hand, an initial shortlist of CROs can be developed based on the project team's understanding of the various CROs' capabilities to deliver on the responsibilities defined.

Table 3.1   Example of a Pro Forma Used to Define Responsibilities of Sponsor, Affiliate Sponsor Company, and CRO

Task/Deliverable	Sponsor	Affiliate	CRO	Elapsed Time
Write protocol	•			Complete
Produce clinical report form (CRF)			•	1 month
Identify investigator site			•	1 month
Ethics committee submissions			•	2 months
Local regulatory approval		•		3 months
Monitor sites			•	12 months
Data entry			•	13 months
Write full report			•	14 months

Gathering Data to Help Shortlist CROs

Information can be gathered informally via industry contacts and colleagues, or from experienced consultants and data services that specialize in this field.

In house. Pharmaceutical companies that outsource regularly should assemble a database of information on CROs that have been used by the company. In this way, staff can make direct contact with others who have used CROs in the past. Such a database is also valuable in analyzing the pattern of CRO use and in helping senior managers identify opportunities to rationalize and leverage CRO use.

Peers and industry contacts. Making inquiries of colleagues in other companies often yields useful insight to CRO performance on particular projects. If used carefully, this informal feedback can be taken from several sources and combined with other intelligence to gain clearer insight into the potential of the various CROs to undertake the project.

CROs. If the project to be outsourced is already well-defined, the project team can establish the experience of a CRO in a given area quickly and use this to initiate a shortlist. This information can be checked formally or informally through referees, publications, and third parties, such as investigators and central clinical laboratories, who have worked with the CRO.

Industry groups. With the increasing use of CROs and the rapid changes in the CRO sector, managers actively engaged in using CROs have formed industry groups to share methods and increase understanding. Groups have formed in the United States and in the UK, and members include representatives from both major pharmaceutical and biotech companies. Such organizations can provide an excellent network for those frequently using CROs.

Consultants. A number of consultancies provide information and resources to identify and evaluate CROs for projects ranging from single studies to long-term strategic relationships. Such support ranges from simply advising a client on the capabilities of particular CROs, to undertaking a detailed operational and financial feasibility evaluation of a group of CROs on behalf of a client. The latter can be performed anonymously if a client requires fresh unbiased information from CROs it has used in the past.

Software and database services are available, as are business-to-business Web-based services. These provide templates to specify the project to be outsourced and to evaluate the impact of different specifications and scenarios on the operational and financial outcomes of the study.

Typically, a combination of these information sources is used to draw up a shortlist of CROs and begin the evaluation process.

Initial Evaluation of CROs

Basic criteria

There are certain basic criteria that a CRO must meet before being considered for detailed evaluation against a planned project. Such criteria include a strong operating history, experienced managers, financial stability, effective project management and quality control procedures, and geographic scope.

Operating history. The CRO must have been in business long enough for its staff to demonstrate their expertise and the coherence of their operating procedures. A minimum infrastructure must also be in place, reflecting the presence of a sufficient number and quality of facilities, systems, and people to deliver on the current work demands. Preferably, some well-known and reputable pharmaceutical companies are amongst the CRO's clients.

Experienced, competent management. Competent and experienced senior managers are essential for a CRO to survive the turbulent contract research environment and deliver a quality product. The management must be able to cope with rapid growth or the consequences of merger or acquisition.

Financial stability. The CRO must demonstrate a steady future business stream, a strong balance sheet, adequate cash flow, and profitability. Many CROs are publicly traded organizations in their own right, or are owned by larger corporations. In these cases, audited financial reports are easily obtainable. However, many CROs are small and young, and more effort and expertise may be needed to assess their financial stability. Because of this range of circumstances and the volatile nature of the CRO sector, an experienced analyst should evaluate the financial status of selected CROs.

Project management. CROs should have systems, processes, and experienced staff who can implement complex projects. They must be able to demonstrate a clear ability to communicate and manage across sites and functions and be able to provide accurate information to the sponsor in a timely fashion. Management responsibility and authority must be clearly delineated.

Quality systems. The sponsor must be confident that the CRO has quality control and quality assurance (QA) systems embedded in its processes and procedures. These must be substantiated by fully documented and validated SOPs and staff training. All of these will be required if projects performed by the CRO are to be credible and acceptable to regulatory authorities. Further confidence can be gained if the CRO has attained external accreditation, such as ISO 9001.

Project-specific factors

If a group of CROs that meets the basic criteria as credible suppliers is identified, the next step is to evaluate the CROs against the particular needs of the project.

A pivotal, proof of principle study in a field that is technically difficult to evaluate and monitor should lead to a shortlist of CROs that have strong experience and expertise in that area. If a project is international in scope, an initial assessment of how the CRO is structured to deliver in all relevant territories is essential. This should address issues such as: where the offices are located, which countries are supported from these offices, and whether these affiliate companies are wholly owned by the CRO or operating as subcontractors. Operating procedures should be truly internationalized and standardized. It is also important for the project to be covered by a single contract that supports a single route of payment in a currency or currencies acceptable to the sponsor. If a fully integrated report is a key deliverable, CROs that can combine safety, data management, and reporting capabilities will be preferred. Table 3.2 suggests basic and project-specific requirements that can help in assessing CROs.

Table 3.2   Basic and Project-Specific Requirements Used to Assess a CRO for Inclusion on the Shortlist

Basic Attributes	Strength (xxx Strong, x Weak)	Comments
Operating history	XXX	Operating for 15 years. Publicly traded company 5 years. Recent major acquisition may divert attention from ongoing projects.
Financial	XXX	Full financial information available; strong balance sheet and cash flow.
Management	XXX	Have managed one major merger and acquisition and taken CRO public.

Project-Specific Attributes	Strength (xxx Strong, x Weak)	Comments
Access to opinion leaders and investigator network	XXX	Recent projects in this therapeutic area indicate strong links with relevant medical community.
Protocol design capabilities	XX	Previous therapeutic area director has taken on other responsibilities—will she be able to dedicate time to protocol development?
Experienced project director	XXX	Project director available due to recently completed study.
Experienced CRAs	XXX	CVs of CRAs indicate several with over 5 years' relevant experience in key geographic areas.
Enhanced patient recruitment	XX	Currently developing more effective links with patient interest groups, associations, Web sites, and so on—uncertain impact on proposed project.
Geographic scope	X	Seasonal project will benefit from CRO being able to operate effectively in both Northern and Southern Hemispheres.
Expedite data collection	XX	Fax-collect, direct data entry. No remote data or Web-based technology.

Having established the basic criteria that should determine the shortlist of suppliers, and utilized good data sources to facilitate the choice, it is appropriate to make a more formal approach to the shortlisted suppliers.

Requesting a Project-Specific Proposal from Shortlisted CROs

The objective at this point is to evaluate the CRO's capabilities for delivering on the particular project—the issues and advantages their feasibility plan shows as compared to the other suppliers and how the costs from each CRO compare.

The key to requesting a detailed feasibility and cost proposal (Request For Proposal or RFP) from a shortlist of suppliers is to ask for *identical* information from each CRO.

The RFP might comprise five items:

- Project protocol describing the objectives, study population, parameters to be measured, analysis and data to be reported, standards and templates to be used, and timelines for delivery;
- Request for details of recent experience relevant to the project;
- Clearly constructed definition of responsibilities. This lists the activities to be undertaken by the CRO, the sponsor company, and any other third party, such as a local affiliate company (see Figure 3.2). The deliverables must also be clearly defined;
- Activity-based request for detail of all staff types, labor hours, and associated hourly costs needed for the CRO to deliver on the project (see Table 3.3). Pass-through costs, such as investigator fees and travel, should be itemized separately; and
- Request for a narrative and supporting data (such as key personnel CVs, relevant SOPs, and quality standards) on how the CRO will construct a project team and deliver the project. This should also request suggestions from the CRO on how delivery can be optimized.

The RFP should be factually correct and compiled by the outsourcing manager with input from the technical experts from the departments involved. Where possible, electronic forms and spreadsheets should be supplied to save transcription.

Selecting the CRO

Many aspects of good procurement practice are considered unnecessary by technical staff in pharmaceutical companies. However, when evaluating and selecting suppliers for key projects, there are benefits to adhering to a framework that is understood by sponsor and suppliers alike:

- The sponsor team must agree up front on the criteria against which bidding CROs will be evaluated.
- The *same* information (RFP) must be sent to all suppliers.
- Delineate the key contact in sponsor company and direct queries through this person. Alternatively, agree that technical experts can be contacted for clarification, but define limits and record discussions. It is important not to impede dialogue and resolution of queries.

Table 3.3 Request for Details of Labor Costs

	Staff (number of full-time equivalents and grade)	Hours	Cost/Hour	Total
Contact investigators				
Site initiation visits				
Monitor sites				
Site closeout visits				
Etc.				

- Do not divulge price or other sensitive information between suppliers; not only does this undermine the credibility of both parties, it is illegal.
- Do not negotiate price, contractual terms, conditions, and so on until the end. Define which staff experienced in this activity will be responsible.
- Apply the same timelines for all suppliers.

When the RFP is sent out, initial queries and requests for clarification from the CRO should be discussed with identified sponsor personnel; a time period should be allowed for this, and any new information that helps clarify the RFP should be sent to all bidders.

The CROs should be directed to return the completed RFP by a set date as a full and final proposal to perform the project.

The sponsor project team should then meet internally to evaluate the proposals. All operational, quality, time, and cost factors should be prioritized and evaluated by the project team. Any queries that the team may have for each of the CROs should be discussed before choosing to meet with the one or two CROs that have provided the most compelling proposal. Such a meeting will comprise detailed discussions on how the project will be delivered and should result in the selection of the CRO to run the project.

The CRO of choice may not provide the lowest price quotation. When CROs respond to the RFP, they should not simply meet the detailed specification, they should describe the means by which they can reduce time to delivery and increase quality. These factors, upon consideration, may have more impact on the true value of the project to the sponsor than small costs above the budget limit.

Prestudy Audit

Most sponsors require a CRO to undergo an audit of their facilities, processes, and procedures to make sure the supplier is able to deliver on the project, to ensure compliance with relevant regulatory standards, and to confirm claims made in the completed proposal. This should not be a process delegated solely to the sponsor's Quality Assurance (QA) Department. Technical experts should become involved in the prospective evaluation of the CRO's processes and practices with the aim of determining how these will support the delivery of the key project outcomes. Experienced staff or consultants should be conferred with to define the methodologies for auditing CROs.

If the selection process is well planned and coordinated, the chosen CRO can be audited prior to commencing the project. The main office or offices that coordinate services and at least one key affiliate office should be visited and audited to understand how activities are managed locally. Some of the key areas that should be evaluated are given in Table 3.4.

The auditing team should determine the particular data to assemble and audit against each item to gain insight and confidence in the CRO's capabilities. A full audit report should be made available to both the sponsor project team and the CRO. It should comprise key findings and recommendations regarding the suitability of the CRO to perform the project. If minor issues are noted for resolution, the CRO should be asked to respond to the recommendations within a certain time period in order for the project to go ahead.

Table 3.4   Examples of Areas to Be Audited When Evaluating a CRO

Area	Key Issues
Offices/Premises	Are these sufficient to house staff and operations? Are the premises secure and can the CRO archive data safely on a long-term basis?
Geographic location	Does the network of offices and local staff seem able to support the current project? How does location affect communication (e.g., time and seasonal differences) and travel costs?
Financial	Establish financial stability and viability through audit of available data. Use staff experienced in this area.
Operating capacity	Can the CRO provide resources for current and anticipated demand?
Staff expertise	Review staff CVs. Assess staff experience against seniority. Are best staff being retained?
Training records	Are staff given appropriate training? Are training records current and complete?
Quality systems and processes	Are Quality Control (QC) and QA systems well managed and integrated? Are operating procedures fully implemented (SOPs) and documented?
Information systems and information technology	Do systems comply with written specifications in terms of reliability accuracy, compatibility with sponsor systems, and so on?
Regulatory	What experience does the CRO have with implementing key regulatory processes (IND, CTX, NDA, MAA, local variations)? Can the CRO demonstrate effective relations with key regulatory authorities?
Data management and analysis	What processes and systems are in place to expedite high-quality data management and analysis?
Report writing	Can regulatory standard reports be generated in a time-effective manner? Can larger regulatory dossiers be compiled? What systems support these activities?
Local effectiveness	Can the CRO demonstrate regulatory experience, local languages, understanding of local medical practice, translation services?
Operating experience	What are the key therapeutic areas in which significant expertise can be demonstrated? What client base is involved? How is expertise designated to projects? Will the project compete with others for such resources?
Project management	How will the project be resources and managed? Do the project and line management processes lend support to a successful outcome? How are resources and progress monitored?
Clinical trial supplies	Can the CRO import/export supplies, package, label, relabel, randomize, and store clinical trial supplies? Can allocation of supplies be managed centrally to optimize utilization of costly supplies?
Support services	Is the CRO able to provide support services on a broader basis such as central clinical laboratory services or preclinical toxicological services?

STAGE TWO: DEVELOPING THE SUPPLIER BASE

As a sponsor gains experience in using CROs, the second stage is to develop a coherent supplier base. This can be achieved by:

- Maintaining confidence in the current CROs by evaluating their performance;
- Ensuring continued adequacy of their systems and facilities through audits;
- Identifying the sponsor's future outsourcing needs;
- Noting gaps in the supplier base; and
- Undertaking a structured evaluation of new suppliers to fill the gaps.

These sets of activities should be run in parallel and are described in the following two sections.

Evaluating Suppliers on an Ongoing Basis

Using metrics to evaluate supplier performance is an important activity. However, it is worth noting that the chosen metrics should:

- Measure what is most important for a particular project;
- Enable the sponsor to plan its longer-term use of the CRO; and
- Be clearly defined and understood by suppliers.

If the sponsor is using CROs extensively across a variety of projects and services, it is important to develop an evaluation system that can work across this varied terrain. This will allow CROs to be compared in a consistent way over a number of projects (Zuckerman 1999). Examples of factors that can form the basis for such a system include:

- Quality (e.g., number of data queries, number of database changes);
- Efficiency and cost (e.g., cost per clean data point);
- Timelines (e.g., first patients recruited, final patient completed); and
- Cycle times (e.g., last observation to database lock).

The extent to which a company should invest time and effort in gathering, analyzing, and utilizing such metrics must be considered carefully in terms of the likely improvement in operating performance. However, the simple step of identifying key outcomes for a project, establishing targets, and measuring delivery against such targets can focus and enhance the performance and collaboration of both supplier and sponsor. Metrics can also be used to construct contractual terms and conditions that reward the CRO for delivery that exceed quality and time targets. Metrics should be supported by a summary of the experience of working with the CRO during a project, written by the project manager. This will help to evaluate performance in such areas as:

- Project management;
- Communication;
- Flexibility;
- Responsiveness;
- Accessibility; and
- Collaboration/team orientation.

By storing CRO profiles, performance metrics, and fees and having such data available to all project teams, CRO shortlists can be assembled rapidly for a given project.

Facilities and systems audits

Care should be taken to remain up-to-date on changes to the systems, processes, procedures, personnel, and facilities of CROs currently in the supplier base. Many of these aspects are checked as part of the project audits performed by the sponsor QA staff. Such checks should also have input from the sponsor's outsourcing group to ensure current operational data. This information, combined with metrics collected from each project, will aim at confirming that the supplier is being managed in a way that ensures the desired quality level.

If a supplier is used extensively or is providing support for critical projects, it is appropriate to undertake periodic (annually or biannually) integrated audits of all systems, process, procedures, facilities, and personnel. A good example of this is a sponsor working exclusively with a single central clinical laboratory. There is a small but critical risk that a systematic error could affect a large number of projects. For this reason, it is appropriate to retain a clear and contemporary understanding of the central clinical laboratory operating standards. Aspects that should be evaluated during such audits overlap considerably with those considered when evaluating a new supplier (see "Prospective evaluation of new suppliers" below).

Identifying Future Outsourcing Needs

Periodically, senior managers should undertake a review of the core/noncore activities of the company and use this review to identify activities that should be outsourced. The current supplier base should be reviewed to determine future fit and then prospective suppliers evaluated to fill gaps identified in the supplier base.

Strategic review

Many companies already have extensive experience working with a number of suppliers to provide a variety of services. The composition of the supplier base should be reviewed against current needs, then evaluated against anticipated needs over the medium to long term (two to five years). Table 3.5 shows a number of factors that can be used to evaluate current and changing resourcing needs.

Identify gaps in the supplier base

This method can be used to identify where current gaps in available services will prevent the current supplier base from meeting future sponsor needs. These gaps should be prioritized and those of most strategic and/or short-term importance should be addressed first.

Prospective evaluation of new suppliers

When gaps in the current supplier base are identified, CROs should be evaluated that seem most capable of meeting the identified needs. This process should occur on an

Table 3.5 Factors Helpful in Assessing Current and Future Resourcing Needs

Factor	Consideration
Core/Noncore	If a company is clear on what its core competencies are, it is much easier to identify what should be outsourced. The more difficult question is, how will this change and how will outsourcing change in order to compensate?
Economic evaluation	This is an extension of the core/noncore issue. A full economic evaluation of the cost of undertaking activities will help identify those that a CRO might perform more effectively.
Operational benchmarking	If metrics on key processes are available, it is possible to benchmark the company versus competitors and CROs. It is probable that CROs can deliver more quickly and at lower total cost in certain areas of operation.
Insufficient/too much internal resource	This is often the key driver of short-term outsourcing. However, if the above strategic factors have been evaluated, what is the optimal balance between internal and external resources in the future?
Geography	If support is required in multiple countries, the advantages of a major international CRO versus two or more local CROs need to be considered.
Scale	Are certain projects of a scale or scope that they cannot be economically resourced internally? Very large clinical trials can be a significant drain on the company's resource and yet might be only of two years' duration.
Technology	Can a CRO provide rapid access to a new technology or technique, either directly or through codevelopment?
Operational focus (e.g., therapeutic)	Is the company going to remain in all current therapeutic or geographic areas? If a new strategic focus is planned, how will an orderly exit from noncore activities be achieved? Can the noncore activities be outsourced?

ongoing basis, separate from the outsourcing of urgent projects. The timetable of events can therefore be more relaxed to allow a full evaluation involving appropriately experienced sponsor staff. Where possible, the same team should be periodically gathered to evaluate the next CRO.

The evaluation process is essentially similar to that used to evaluate CROs for individual projects (see "Initial Evaluation of CROs" earlier in this chapter). However, the focus of the evaluation is likely to be broader since the objective is to use the CRO in a greater number of projects and (probably) to utilize a wider variety of the CRO's services and geographic locations.

The sponsor team should request comprehensive data on facilities, staff, systems, project experience, and so on for review before undertaking a more formal, on-site review of capabilities. The on-site review should involve meetings and presentations from both CRO and sponsor staff and include affiliate company staff.

At the end of the review process, the sponsor team should assemble a comprehensive report recommending whether or not to contract with the supplier, either on a broad- or limited-service basis in the first instance.

STAGE THREE: DEVELOPING LONG-TERM PARTNERSHIPS

Through experience in cooperating with a group of CROs, a small number will emerge as preferred in terms of ease of interaction, quality and scope of provided service, and fit with likely future outsourcing needs. There can be important benefits in applying more effort to develop such partnerships for the long term (Doz and Hamel 1998)—this is the third stage of working with CROs.

How Does Partnering Differ from Other Relationships?

An organization that outsources most of its activities, such as a virtual drug development company or a company that has extensive experience working with CROs (i.e., is at stage two), will identify the need to maximize the value derived from each of its suppliers. To do this, more effort must be focused on these relationships, including:

- Management time (e.g., joint management team and a dedicated relationship manager);
- Sharing information to support joint strategic decisions (e.g., CRO may invest in fixed assets or more staff in response to anticipated need of sponsor); and/or
- Systems alignment (e.g., sponsor has access to real-time project management data).

In return for applying such effort, benefits should begin to accrue to both partners in the areas and forms outlined next.

Communication. Regular and open communication will highlight mutual needs and expectations, resulting in greater commitment to important decisions.

Planning. Sharing of information regarding future projects allows the CRO to resource projects more effectively.

Organizational knowledge. Greater mutual understanding of procedures and routines will significantly hasten contracting, initiating, executing, and completing projects. Quality is likely to be higher as well.

Empowerment. Greater sponsor trust in a CRO will allow more extensive sharing of responsibilities. This, in turn, should allow the CRO to operate more efficiently, permitting the CRO to be more flexible in its response to changing sponsor needs.

Economics. If a sponsor is outsourcing all of a particular activity to only one partner, the CRO will reap volume benefits, which should translate into cost and time efficiencies that can be shared with the sponsor.

Commitment. As confidence and trust between the partners build, a partnership will develop. This may translate into a commitment on the part of the sponsor to outsource all of a certain class of work to the CRO, with an appropriate lead time given to the CRO to find resources for such projects. Such a commitment is based on the CRO's desire to provide consistently improving services and the sponsor's desire to place return business with the CRO as a result. Such mutual commitment is difficult to encapsulate in a written contract. Indeed, creating a sponsor obligation to place all work with a CRO could lead to complacency in the CRO. A sponsor commitment to reach a certain spending target will lead to a focus on cash spent rather than on the best projects to outsource. A partnership should be based on a commitment to develop excellent outcomes and not on meeting volume or cash targets.

A number of scenarios exist in which it will be counterproductive to partner:

- **Immature relationship.** A partnership is most likely to develop from a relationship that has evolved though stages one and two. It is unlikely that a relationship with a freshly chosen CRO will operate as a partnership unless there is a very strong and immediate mutual reliance.
- **Verifying access to best practice.** It is important to evaluate a spectrum of CROs over time to ensure that current partners are leading suppliers. This reinforces the need to develop suppliers through stage two before reducing choice.
- **Lack of long-term mutual benefits.** The mutual commitment to partnership depends on long-term mutual benefits. If benefits cannot be readily identified and measured, there is little point in committing the effort to a relationship over the long term. The balance of benefits must also be equitable for the partnership to be sustainable.

Selecting CROs for Partnering

It is most likely that a strategic partner will be selected following a period of good experience in stage two. It is unlikely that a top-down imposition of a partnership by senior management will garner approval and support from the primary customers—usually the project teams. Customers must consent to the concept and should drive it. Project teams, in turn, will have to accept some loss of control and will have to learn new skills in managing third parties as part of the project team structure.

Final selection of a partner should occur only after the CRO is identified as:

- Meeting a particular medium- to long-term need;
- Operating according to a strategic direction complementary to the sponsor's (e.g., the CRO is building particular strength in a therapeutic, geographic, technical, or informational area that is needed by the sponsor); and
- Possessing qualities that, over time, have demonstrated the existence of a good cultural fit that can be developed effectively.

Partnering should be used as an opportunity for the sponsor to focus on its core activities and engage with partners having complementary skills. This will ensure greater mutual interest in the partnership and should enable both partners to extract maximum economic benefit.

Evaluating a Potential CRO Partner

Partnering is an area that is still novel to the industry. Partners are, by definition, dissimilar organizations. Typically, each organization experiences staff changes at all levels and has few objectives fixed over the medium to long term. Such ambiguity requires that the partnering relationship be allowed to evolve. The key is to ensure that the partnership is positioned within a clear strategic framework and not judged on short-term microobjectives.

To prevent such judgments, some quick wins should be set up to highlight benefits accrued from each partnership. Senior managers, project, and line staff from both partners should establish interface, which will require a joint effort to ensure that direction and outcomes are meeting real business needs. Metrics should be established to evaluate the contributions of *both* partners and to link future relationship path to successful results, allowing a relationship to grow on the basis of value created for both partners.

CONCLUSIONS

This chapter has provided a set of tools to effectively identify, evaluate, and select CROs and a framework within which to develop an appropriate supplier base. Each pharmaceutical company and CRO will have accumulated different experience and will have been in relationships that have reached different stages of development. The tools and approaches used to identify and evaluate CROs should be adjusted accordingly. By working through stages one through three, a significant outsourcing capability can be developed to support both immediate and long-term business needs.

REFERENCES

Doz, Y. L. and G. Hamel. 1998. *Alliance Advantage: The art of creating value through partnering.* Boston: Harvard Business School Press.

Quinn, J. B. and F. G. Hilmer. 1994. Strategic outsourcing. *Sloane Management Review* 35(4): 43–55.

Zuckerman, D. S. 1999. How to best measure the performance of clinical trials. *Scrip Magazine,* October. 22–23.

Nadia Turner and Chris Keep

Financial and Organizational Outsourcing Considerations

Assuming that the majority of the authors and readers of this book belong to organizations whose primary objective is to increase shareholder value, it is perhaps not surprising that organizational and financial factors come to the fore when considering the relative merits of outsourcing options.

Before embarking on our review of the financial and organizational considerations, we feel it appropriate to state, allbeit obviously, that any outsourcing decision should be made within the context of an integrated sourcing strategy, the objective of which should be to obtain maximum value from both internal and external skills, knowledge, and technologies. If we look at the benefits sought from such a strategy, we begin to touch on some of the issues that we will examine in this chapter. The benefits sought include:

- Maximizing returns on internal resources/technology by concentrating investments and energies on core competences/capabilities for the future;
- Leveraging external skills and investments to maximize value and reduce risk; and
- Increasing the organization's flexibility and responsiveness in a turbulent environment.

Integral to the successful development and implementation of an effective sourcing strategy is a clear and rational process and associated responsibilities for the make/build versus buy/disinvest decision at both a strategic capability level and the individual project level. The decision process will need to take into account the following criteria:

- The impact of the activity on value and risk;
- The organization's ability to perform the activity (including a benchmarked economic analysis); and
- The activity's long-term strategic priority.

Sequentially assessing an activity against these criteria should result in a clear, objective indication as to whether the activity should, in the long term, be outsourced as policy or maintained/built internally. However, notwithstanding the long-term policy, the reality of the resource demand from drug development programs is such that there will be occasions where activities, which by policy are to be conducted internally, may require an outsource solution. Consequently, there will be a number of potential outsource scenarios, each with different objectives and, therefore, different financial and organizational considerations. We will examine these later in the chapter. We will first explore further the financial and organizational factors that come to play.

FINANCIAL CONSIDERATIONS

General

Given the increasing quantity and value outsourced by pharmaceutical companies to Contract Research Organizations (CROs), financial considerations must become more important (see Figure 4.1). The financial considerations of outsourcing depend on several factors:

- The nature of the service being outsourced;
- The type of supplier(s) considered;
- The difficulty of sourcing or purchasing such services externally; and
- The relationship scenario between the pharmaceutical company and CRO.

The most obvious financial consideration is the price paid to the CRO. However, this needs to be considered in context with other factors, such as:

Figure 4.1   Difficulty of purchasing externally versus type of service and supplier

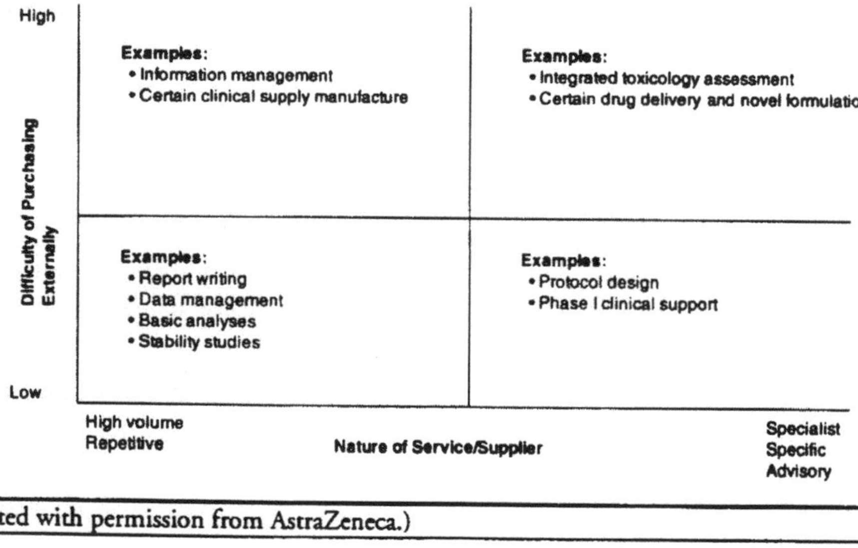

(Printed with permission from AstraZeneca.)

- The deliverable's fitness for purpose/appropriate quality;
- The financial transparency of the cost structure; and
- Any incremental internal costs.

Market circumstances for the CROs may affect price over time. Pharmaceutical company factors affecting price are often driven by time and capacity/availability considerations.

Typically, price appears as a second- or third-level criterion. This does not mean that price transparency and detail are not important in ensuring that price is understood in terms of ensuring value for money; indeed, price may be a differentiator when higher-priority factors are closely balanced.

Price Transparency and Cost of Activities

Pharmaceutical companies carry out in house the same or similar activities as those outsourced and, for a big pharmaceutical company, the volume of work done internally may be on a similar scale to the volume done by a CRO. Particularly in the case of the more standard or repetitive activities, a pharmaceutical company could use its knowledge and experience to develop models against which the CRO costs can be compared. This can be a revealing exercise, but it may require significant effort initially within the pharmaceutical company and may necessitate an uneasy agenda in discussions with CROs. Pharmaceutical companies may not have such information readily available, requiring manual calculations, estimations, and some assumptions. A CRO's revenue comes from billable hours to clients, while a pharmaceutical company's revenue comes from drug sales. The CRO's areas of activity are profit centers, while in the pharmaceutical company, these are cost centers. The emphasis or need to measure the total real cost of activities done in pharmaceutical compaies is different.

There is a difference between clinical and nonclinical activities in terms of price transparency. Traditionally, CROs provide more detail for clinical than for nonclinical activities. However, the information is available for both since this information is key for the CRO to understand the economics of its activities. On the other hand, there is a historical issue of pharmaceutical companies not examining nonclinical costs in the same detail as they examine clinical costs.

Cost Bundling

As Figure 4.2 demonstrates, in selecting a CRO, a pharmaceutical company must assess what type of service or supply it needs. Bundled costs for different subactivities need to be broken down to ensure that this does not represent a financial disadvantage. Being charged a single rate across a range of activities may mean lower-level operational activities being charged at a rate for higher-level tasks. The price and volume metrics of the work must have enough transparency to allow comparisons to be made.

Price and Quality

There is a danger when price is not given enough focus because of concerns about quality being compromised. Quality must be defined and will include criteria such as:

Figure 4.2   Need versus type of service and supplier

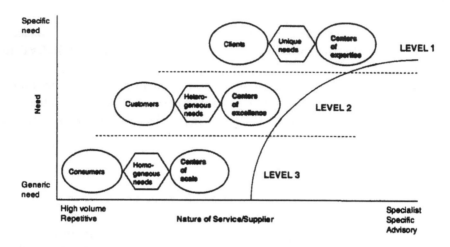

(Printed with permission from AstraZeneca.)

- Availability of CRO resources;
- Capacity—geographic capacity, specific locations;
- Expertise—specific expertise, technical or operational;
- Continuity with previous and future projects or staff; and/or
- Flexibility and adaptability—responsiveness to pharmaceutical company needs, changing or differing requirements.

It can be argued that quality appropriate to the task should be a price of entry criteria. There may be levels of quality in aspects of the work performed by the CRO that can add value to the basic requirements of the task, but these quality levels must be understood and mutually agreed upon, since there are usually financial consequences that accompany them.

A metrics process must be used, such as a dashboard/balanced scorecard to monitor quality factors and ensure that the appropriate balance between them and price is achieved. Metrics are a subject in their own right, but it should be noted that the number of metrics, their complexity, and the effort required for data collection must be kept low to ensure success in implementation and continued use.

Price and Pharmaceutical Company/CRO Relationship

Price is a factor for all relationships between the pharmaceutical company and the CRO, from the occasional transactional through to the regular or continuous in frequency, and performance improvement or value-added in nature.

Much work may have gone into defining working processes and relationship management structures, tailoring the CRO's services to the pharmaceutical company's needs. These additional CRO activities can translate into costs somewhere and possibly higher rates.

The pharmaceutical company and the CRO should jointly develop and use cost models and dashboard metrics to bring transparency to the relationship. This is important in ensuring that the cost of the value-added aspects represents what was expected and value for money from the relationship. Transparency in the relationship will avoid awkward discussions at relationship-review meetings, where such issues are likely to be revisited.

Total Cost of Outsourcing

The price paid to the CRO represents a major portion but not the total cost of outsourcing (see Figure 4.3). There are internal/incremental costs for:

- Relationship management;
- Project management;
- Interfaces; and
- Changes, performance deviations, and late delivery.

Analysis of one year's outsourcing at a global level across clinical and nonclinical activities indicated incremental costs of 12 percent to 15 percent. For a big pharmaceutical company, this can represent $10–$30 million U.S.

Relationship Management

The incremental costs of relationship management are mainly determined by the number of CRO suppliers and the way the outsourced activities are distributed across them. Reduction of the tail of the supplier base is possible by focusing standard activities or bundles of work with fewer preferred suppliers. This may allow the chosen CROs to reduce costs due to greater forecast volume with continuity, making investment in process improvements to the relationship possible and worthwhile. There will always be a tail to the supplier base, but this is where only the specialist/niche suppliers should

Figure 4.3   Total cost of using a CRO

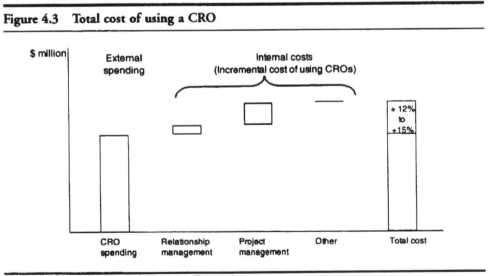

(Printed with permission from AstraZeneca.)

exist. Having fewer but larger CRO relationships for the same volume of outsourced activities may mean that incremental costs are reduced or more value is gained from similar costs. Estimates indicate that ad hoc CRO supplier relationships may have more than double the incremental costs associated with them than preferred CRO supplier relationships over the longer term (see Figure 4.4).

ORGANIZATIONAL CONSIDERATIONS

The importance of leadership and clarity of responsibilities cannot be overestimated when considering outsourcing. If the organization is to stand any chance of realizing the benefits of outsourcing as a strategic weapon, a sourcing strategy must be developed and implemented, which requires clear leadership and drive at senior management level. There must also be clarity as to who makes the key sourcing decisions at both a capability level and project level (i.e., the "make versus buy" decision and the selection of supplier). In most organizations, the discussions about who is responsible will tend to involve three key players: the project manager, the resource manager, and the purchasing or procurement manager. Thereafter, supplier management will also be the concern of these three players. Last but not least, the responsibility for sourcing process ownership must be clearly defined whether at the project, departmental, or corporate level.

Clearly, the relative roles of the project, resource, and purchasing managers will vary across organizations, largely depending on the budget-holding responsibilities of these players. However, we have found it useful to consider the following as a model that describes three distinct roles:

- **Project manager**—responsible for managing the project deliverables through the supplier;
- **Resource manager**—usually a technical line manager, responsible for monitoring the technical quality of the supplier's process; and

Figure 4.4   Supplier base use and incremental costs

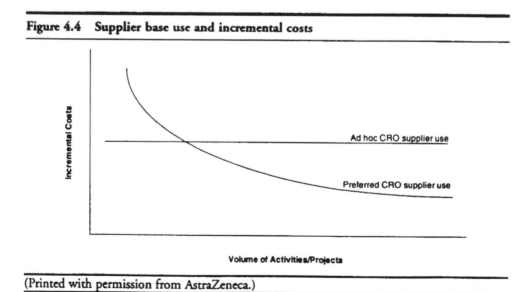

(Printed with permission from AstraZeneca.)

- **Relationship manager**—responsible for negotiation of terms and, thereafter, accountable for managing the customer relationship with the supplier across projects, for performance review, and for ensuring that the supplier continues to be "fit for purpose" in the long term (e.g., through investment).

We believe that this model can serve as a practical blueprint when considering organizational structures to manage outsourcing. Implementation of this model can be aided significantly by the completion of a RASCI (Responsible-Accountable-Supports-Consulted-Informed) chart that defines the key tasks/decisions associated with the outsourcing process, as well as the respective responsibilities of the various interested organizational groups, departments, and individuals. An example might include the subset of tasks and responsibilities shown in Table 4.1.

Clearly, a key part of the process is the "make versus buy" decision, and understanding who has ultimate responsibility for this decision is essential. In most organizations, this decision falls to either the project team or the technical department, which would normally supply the resource if the task were done internally (organizations tend to be either function led or project led and this will undoubtedly determine the responsibility here). The distribution of the RASCI will also vary according to the nature of the client/supplier relationship (i.e., whether it is a strategic function-wide or departmental tactical supplier). In any case, it would be important to ensure that the RASCI chart is not only constructed and supported by all key players in the outsourcing process but also that it is used as a working model to ensure clarity and consistency in operating the process.

Outsourcing Scenarios

As mentioned in the introduction, we have found it useful to define a number of outsourcing relationships with differing objectives. These are categorized according to the frequency of use of the supplier and the potential for the relationship and are summarized in Figure 4.5.

Organizational factors will differ according to the relationship objective. For example, an occasional transactional relationship whose objective is to evaluate the supplier and/or to use the supplier occasionally for capacity only would be procedure-focused

Table 4.1   Simple RASCI Chart

	Purchasing/ Commercial Expert	Project Team	Resource/Technical Manager
"Make versus buy" decision	S	C	AR
Select supplier	S	I	AR
Commercial/contractual negotiation	R	I	A
Manage project deliverables	I	AR	S
Manage and develop relationship	R	I	AR
Manage and develop technical quality	I	I	AR

(Printed with permission from Astra Zeneca.)

Figure 4.5   Defining outsourcing relationships

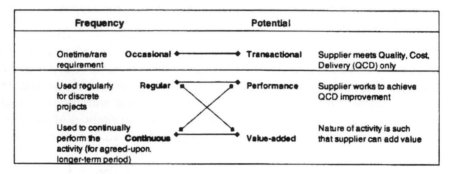

Note: All relationships start as occasional/transactional.

(Printed with permission from AstraZeneca.)

and driven by efficient standard operating procedures (SOPs), whereas a continuous value-adding relationship would involve the setting up of integrated client/supplier project teams with additional relationship management by a specialist sourcing group whose responsibility would be to develop and manage the outsourcing process and to manage strategic supplier relationships. Likewise, on a financial note, the long-term basis of pricing will also vary. An occasional transactional relationship would be based on a competitive tendering process whereby a continuous value-adding relationship would require best-in-class benchmarking of both performance and performance improvement.

Paul Ranson

Clinical Research Contracting—
A Legal Perspective

Self-evidently, national practices, laws, and language will lead to local variations. Particularly in the United States, the Food and Drug Administration (FDA) requirements will, for instance, result in provisions that are not found in the European Union (EU). Heightened legal sensitivity in the United States to issues of liability and litigation risks in general results in somewhat more detailed contractual documentation than would normally be seen in Europe. In addition, it has been in the United States that we see glimmerings of a greater degree of risk sharing between CROs and sponsors. To the extent these relationships involve technology transfer, wholly new considerations, such as the licensing of patents and know-how, will come into play.

"Contract research" is a term that covers a multiplicity of tasks and levels of service. For instance, it may denote simply a single, discrete function, such as data analysis. Conversely, it may imply the performance of every aspect of a clinical research project. For the sake of comprehensiveness, the following review assumes a full-service arrangement.

Even with the disparities discussed previously, the 1997 International Conference on Harmonization (ICH) Good Clinical Practice (GCP) Guidelines result in greater uniformity in terminology and format. Despite obvious local jurisdictional and practical differences and its absence of any significant degree of EU contract law harmonization, it is possible to offer general guidance and insight on the current provisions commonly seen in contract research agreements.

This chapter reviews the following issues:

- General legal considerations in any outsourcing arrangement;
- Preparation and contractual structure;

- Precontractual documentation, including confidentiality agreements and letters of intent;
- Clauses commonly found in a contract research agreement;
- Managing change; and
- Employment issues.

GENERAL LEGAL ISSUES IN OUTSOURCING

Outsourcing is, in many ways, an extraordinary commercial arrangement.

Trust. Outsourcing puts the control of a sponsor's commercial destiny in the hands of a complete outsider. By the nature of the relationship, there can be no direct control over the CRO.

The CRO will be acting on the basis of payment for services provided, expenses, and profit. To some extent, the CRO's interests are in conflict, from the outset, with those of the sponsor. That conflict must be controlled so that both sides benefit without exploitation of either of them. Otherwise, the outsourcing arrangement will lead to trouble.

The sponsor puts itself in a position of having to place a high degree of trust in the CRO to:

- Act fairly;
- Act in the sponsor's interest; and
- Perform to the level that could be expected of the sponsor's own direct employees.

The law has not caught up with outsourcing as a commercial practice. The law recognizes no special relationship with or obligation of the CRO comparable to those that might exist between a company director and a company, or a lawyer and a client.

Legal protection for the sponsor arises only from:

- What is built into the contract; and
- How the outsourcing arrangements are structured.

Dependency. It is easy to see how the sponsor could be in a weaker position than the CRO, since the sponsor's business-critical information and systems, together with direct control over and direct access to them, pass from a sponsor executive to an outsider. Control and access are no longer through the sponsor's own employees under direct executive control.

Whole or partial loss of access and control can have serious operational implications. In an extreme case, where the CRO ceases trading through insolvency, the sponsor's business could be severely affected.

Disentangling the relationship. What if things go wrong or circumstances change? The sponsor may be in a dangerously weak position if the contract does not deal adequately with the return of resources and materials from the CRO on termination.

The sponsor may be able to terminate the contract if the arrangements are not operating satisfactorily. It must be able to bring its contract research function back in house speedily or transfer it to another CRO, in circumstances where the new CRO may not be inclined to be cooperative or may be unable to cooperate.

The role of the contract. The onus is on the sponsor to ensure its interests are protected in the contract. The CRO is obliged to do what is specified in the contract, but nothing more. The sponsor must negotiate and demand to include any provisions needed for its own protection or, preferably, to use the sponsor's own contract draft.

The sponsor must also take the internal steps necessary to manage the contract and the relationship, to monitor developments, and to have in place contingency arrangements when the unexpected happens.

The positive commercial importance of the contract and contracting process must be recognized and the contract should be viewed and used as a crucial management tool. It should, for both CRO and the sponsor:

- Specify accountabilities;
- Set aims and objectives;
- Specify good performance (standards, quality); and
- Establish regular, planned performance monitoring.

The process of negotiating the written contract is a vital opportunity to improve the prospects for a successful implementation and continuing good relationship by:

- Clarifying prior understandings and negotiations by committing them to paper in a logical, structured way for each party to see;
- Ensuring both sides really do understand each other and what is expected of them;
- Dispelling any prior misconceptions; and
- Reviewing precontractual planning.

A good contract should cover all of the major aspects of the future relationship with clarity and with sufficient detail. To set out commercial understandings clearly, unambiguously, and in sufficient detail requires good analysis and good communication skills. Good, clear commercial agreements improve the prospects considerably for avoiding disputes and successfully implementing the contract.

Ideally, the final, written contract should act as a binding commitment that marks commencement of the long-term relationship.

PREPARATION FOR OUTSOURCING AND CONTRACTUAL STRUCTURE

Defining Requirements

First, the sponsor must analyze its contract research requirements and record them with sufficient detail and clarity. The sponsor will incur additional costs if it finds that it requires the CRO to perform further tasks and these other activities are not clearly covered by the contract and included in the contract price.

The sponsor, therefore, needs to decide the level of detail in which to specify its requirements. For large organizations in particular, the task can be time-consuming, creating a temptation to cut corners. However, expenditures on this analysis at the outset should be recouped handsomely in money and disputes avoided later.

In addition to analyzing contract research needs, future changes to those requirements must be considered and provided for in the contract.

Once the necessary contract research services are identified and specified, defining the performance required from the outside organization in plain, clear, and unambiguous terms is critical. Performance standards are indispensable. How much of each particular service must be provided? How fast are services to be provided? When are they to be provided and how often?

An outsourcing project team appropriate to the scope, complexity, size, and sensitivity of the proposal must be created. During outsourcing procurement, the sponsor must be prepared to invest consultancy and legal support resources. In addition, the project team will have to deal with issues where the function was previously performed in house. Staff morale may well be poor, as employees face the uncertain future of being transferred to a contractor.

Tendering

It is increasingly common for contract research to be tendered. The sponsor should begin by eliciting expressions of interest. Next, the sponsor should issue an Invitation to Tender (ITT), containing an overview of the sponsor's tendering instructions and terms, requirements, and a draft contract, to a number of potential CROs. The issue of the ITT should be under strict terms of confidentiality. Also, where appropriate, the sponsor should include basic information about the assets and services that the potential new outsourcer will inherit. The bidders should be encouraged to respond to the ITT in the prescribed form, thus allowing the sponsor to concentrate on the technical and commercial differences between the bids. Because reviewing the bidders' responses is very time-consuming, the sponsor must plan to reserve the necessary resources. If the size of the outsourcing project makes it appropriate, a data room should be set up.

Even in the absence of multiple ITTs (where a sole bidder is identified as appropriate), preparing a memorandum detailing the sponsor's objectives and the precise functions to be outsourced is beneficial because it forces the sponsor to:

- Clarify what it expects the CRO to do; and
- Clarify the services/work that the sponsor is required to perform.

Multiple bidders generate a more competitive environment. A sponsor may thus use the negotiation phase as an evaluation tool.

Both parties will be looking for financial stability. Management styles and business languages should be compatible, since cooperation is essential to outsourcing success. A one-sided contract will benefit neither the CRO nor the sponsor in the long term, since, if all goes well, this will be an ongoing relationship.

Due Diligence

The sponsor will wish to perform a due diligence assessment of the CRO. Master contract research agreements are, by their nature, long-term, requiring a significant commitment from the sponsor. This is particularly the case if the sponsor is transferring its assets and staff to the CRO. Accordingly, each party will want to satisfy itself that the other is financially sound and can properly perform the terms of the agreement. If the CRO goes into liquidation, it could leave the sponsor without critical data, making trials incomplete. If the sponsor fails, the CRO will incur losses arising from the unreimbursed expenses of setting up programs, acquiring equipment, and taking on staff.

At an early stage, the sponsor will also need to see if the intended CRO is capable of performing to ICH GCP and other required standards, and this will require the sponsor to conduct detailed site and facility inspections, and to obtain and verify other client references.

Contractual Structure

Difficulties in achieving a formal written agreement prior to the start of a trial on a tight timetable, coupled with the problems inherent in letters of intent (see below), have made master agreements increasingly popular, particularly in the United States. Under a master agreement, the ongoing relationship between a sponsor and a CRO is governed by an agreement covering all potential projects. Each specific project then requires negotiation only with regard to commercial terms and specific performance criteria, without continually needing to revisit such thorny problems as liability, indemnity, and other potentially troublesome clauses. Master agreements may or may not involve the CRO in continuing obligations between the projects, which might entail fixed payments or minimum levels of business.

Another frequent practice is the distinction made between the legal provisions and the precise specification of the services to be provided. Ideally, the detailed statement of the services (commonly referred to as a service level agreement or SLA) is attached to the legal documentation as an ancillary document. The SLA should be negotiated by the technical staff, but should be reviewed by the lawyers to ensure general compatibility with the legal agreement.

PRECONTRACTUAL DOCUMENTATION

Legal Commitment and Letters of Intent

Invariably, however, local law will require a degree of certainty before it will regard an agreement as a binding contract, and it may not be binding if it is too vague or incomplete. In addition, where the parties state that the agreement is subject to contract, this suggests that it is incomplete until the details of the formal contract have been settled and approved by the parties. However, the fact that the parties envision that the letter will be superseded by a more formal, contractual document may not, in itself, prevent the earlier document from taking effect as a binding contract. Similarly, an agreement to agree—where the parties have simply agreed to negotiate—may well be too uncertain to have any binding force. For example, this type of agreement may not effectively impose any obligation to negotiate, to use best endeavors to reach agreement, or to accept reasonable proposals.

The question of whether a binding agreement has been formed also arises with a letter of intent, whereby one party indicates to another that it is very likely to place a contract with that party. Negotiations can extend beyond the required project start-ing date, and the parties may start the project based on a letter of intent or on an agree-ment that is subject to contract. More often than not, the letter of intent is worded so as not to create any obligation on either side. However, it is important that this is made clear and that such letters do not contain, for example, an invitation to com-mence preliminary work, which might arguably create an obligation to pay for that work in any event.

Whether an agreement is binding may be determined by national laws. The words "subject to contract" are normally intended to negate contractual intention, so that the parties are not bound until formal contracts are exchanged. However, a letter of intent or letter of comfort may actually be deemed evidence of contractual intention, and it is for the courts to decide whether the parties are bound by the agreement. Courts may be more inclined to deem an agreement binding where the parties have acted on the agreement for a long period of time, or incurred expenditures or time costs in relying on it.

Confidentiality Agreements

During negotiations, it is inevitable that either or both parties will need to disclose to the other their own proprietary or confidential information to enable the recipient to evaluate its interest in a prospective outsourcing relationship. If such confidential information is to be kept secret, a confidentiality agreement should be entered into before that information is disclosed. Furthermore, the disclosing party should consider its overall approach to protecting confidential data against misuse. The confidentiality agreement should be but one part of an overall confidential information management strategy.

Before even contemplating the disclosure of confidential information as part of negotiations, a company should consider the level of trust it has in the other negotiat-ing party. As with any litigation, suing a prospective trading partner for infringing on a confidentiality obligation is a big step and proceedings are extensive and expensive. Most of the damage is irreparable, practically speaking, when confidentiality has been broken; suing the party in default rarely provides adequate compensation. Where there are *any* concerns that the secrecy of confidential information may be violated, it may be better to walk away at that stage!

It is important to take all the time necessary to make sure the confidentiality agree-ment is complete, clear, and properly binding. There is often tremendous pressure to put a confidential disclosure agreement in place quickly, but it is frequently the pre-cursor to more complete agreements, and the precedents set out in the preliminary agreement, such as agreeing governing law, may later be difficult to revisit in the main agreement.

It is therefore advisable to have a full set of agreed-upon standard terms and condi-tions or full contracts that can be adapted for use at short notice. These agreements could cover at least the following situations:

- Disclosure of confidential information;
- Receipt of confidential information; and
- Two-way disclosure.

COMMON TERMS IN A CONTRACT RESEARCH AGREEMENT

A contract research agreement will commonly include the following provisions, which can be grouped into six principal categories:

- Performance terms;
- Financial or commercial terms;
- Data ownership, confidentiality, intellectual property rights, and publication;
- Warranties, indemnities, compensation, and insurance;
- Termination arrangements and consequences; and
- Other legal clauses (boilerplating).

Performance Terms

Performance terms tend to consist of two types. First, broad provisions cover the quality of the CRO's performance, including:

- Conformity with applicable laws (ICH GCP), codes, highest industry standards, and Standard Operating Procedures (SOPs);
- Quality of the CRO's assigned staff are approved for the replacement of key staff members; and
- Reporting and liaison between the sponsor and the CRO.

A sponsor may wish to ensure that certain general standards of diligence are met. Requirements for the use of "reasonable efforts," "best endeavors," or "commercial best efforts" are thus imposed.

Second, specific provisions cover the sponsor's specific requirements regarding the services sought from the CRO, together with, if appropriate, performance criteria, estimates, and whether or not the cost is included in the overall contract price. Such terms will either be in the body of the agreement or referred to in an attachment (i.e., the SLA). Again, assuming a full service contract, the agreement may require CRO involvement in the following tasks:

- Application for study approval to both regulatory authorities and independent ethics committees and maintenance of the approval;
- Preparation and dissemination of protocols, Case Report Forms (CRFs), patient diaries, and so on;
- Packaging, distribution, and randomization of drug supplies (study drug, control drug, and/or placebo);
- Prestudy scrutiny of investigators and sites;
- Organizing and conducting investigator meetings;
- Routine monitoring of sites and site management;
- Site Quality Assurance (QA) audits;
- Adverse event reporting;
- Project management;
- Data entry, review, programming, and transfer (particularly taking into account the risks of clinical trial fraud);
- Statistical analysis;

- Preparation for the final report;
- Trial completion visits to sites; and
- Regulatory submissions.

Financial or Commercial Terms

There has been a move away from the fixed fee structure to one based on the performance of individual units based, for instance, on the tasks listed above. This gives the sponsor increased control over the escalation of costs and limits the potential for unpleasant surprises.

Special attention must be given to unforeseen costs, for example, those resulting from a change in the scope of the contract and the payment of penalties for delayed performance. While contract law may limit the enforceability of such penalties, sponsors are more frequently seeking such extra sums to cover resulting losses. In addition, many agreements will involve performance incentives—an elementary example of a risk-sharing arrangement.

When a project falls behind schedule, agreements commonly propose a mechanism for seeking a solution to the issue. Inevitably, this requires some flexibility on both sides, but may entitle the sponsor to take unilateral remedial action if it considers it necessary. Sponsors would normally confirm that any concessions made would be without prejudice to any other rights under the agreement, including penalties for delayed performance and early termination.

Invariably, the routine details of invoicing frequency, payment terms, and even arrangements for resolving any invoicing disputes will be provided for, as will responsibility for payments to third parties, including investigators and their reimbursement.

Data Ownership, Confidentiality, Intellectual Property Rights (IPRs), and Publication

The agreement will confirm the sponsor's ownership of all records relating to the project, including CRFs, and will impose minimum retention periods in accordance with legal and ICH GCP requirements. The return of materials on termination of the contract should also be addressed.

The CRO is usually required to accept that confidential information given to it, or otherwise acquired by it during the course of the project, is the sponsor's property and must be kept confidential. The agreement may also require that the CRO impose equivalent obligations on its staff and investigators. "Confidential information" should be defined in detail sufficient enough that the parties are aware of what is and what is not confidential. The agreement will usually provide that all confidential materials be returned to the sponsor at the end of the trial or the termination of the agreement and that the confidentiality obligation is intended to survive termination of the agreement.

While IPRs are generally unlikely to arise, a typical CRO agreement confirms that the sponsor shall own all IPRs relating to the drug and the project, arising prior to, or during, the course of the trial. The CRO is usually required to enter into a formal assignment of such rights, as and when required. Typically, the CRO is obliged to disclose any inventions, know-how, copyright materials, and so on, and the sponsor will have a right of access to all data and information, as well as unrestricted rights to use them. The CRO may be obliged to impose back-to-back obligations on investigators, so that IPRs generated by the investigator become the property of the sponsor. The CRO will wish to retain those IPRs it regards as its proprietary methodology.

The sponsor will want to reserve control over the timing and content of any publications relating to the project. In particular, treatment of publications on multicenter studies would need to be addressed.

Warranties, Indemnities, Compensation, and Insurance

Investigators and institutions have long expected to receive an indemnity against claims not arising through their negligence. The ICH GCP guidelines provide that (if required by the applicable regulatory requirements) a sponsor is obliged to provide an indemnity for the investigator, except for claims resulting from the investigator's malpractice and/or negligence. A standard provision in a contract research agreement provides for an indemnity to be given by the sponsor to the institution/investigator against any third party claims for injuries or death caused by the use of the sponsor's product in the trial, provided that the institution/investigator has not been negligent, has followed the protocol, and has notified the sponsor immediately of claims. It is usual to provide that the sponsor is allowed to take over the conduct of the defense of any claim, and that the institution/investigator will cooperate in the defense. Since the investigator is not a party to the contract research agreement, the sponsor may issue a separate letter of indemnity to the institution/investigator, commonly as part of the trial documentation.

As far as the CRO is concerned, an agreement will usually provide that the sponsor indemnifies the CRO against any claims made by third parties for injuries and damage caused by the clinical investigation of the sponsor's product, except in the case of negligence, misconduct, or breach of contract by the CRO. Again, this indemnity will be subject to the CRO's giving prompt notice of any such claim, passing control to the sponsor, and assisting it in the defense. Typically, an agreement provides that the CRO will not settle or compromise such a claim without the sponsor's prior written consent.

The sponsor may wish to impose an agreement provision stating that the CRO will indemnify the sponsor where liability results from any negligence, act of omission, or breach of contract by the CRO, its employees, or agents.

Indemnity is clearly related to the issue of warranties—contractual promises by one or both parties that they will properly perform key aspects of the contract.

In most agreements, both parties will wish to exclude liability for consequential (indirect) loss, liability for loss of profits, and the like. Given that consequential loss is not a legally recognized term in every jurisdiction, it may be appropriate to define in full what liability is to be excluded (for example, subsequent postmarketing personal injury claims).

The agreement may also address the question of insurance. It may provide that one or both parties are obliged to purchase and maintain fully comprehensive insurance policies against all claims that may arise during the contract period. The CRO is normally expected to have "errors and omissions" or equivalent insurance along with other public liability coverage, up to a satisfactory amount, and the sponsor may wish the

CRO to confirm or provide evidence of the extent of coverage. The CRO may also be expected to ensure that institutions/investigators themselves are adequately covered.

Publication is becoming an increasingly controversial area. Nowadays, investigators will not wish to be constrained from publishing, and a sponsor will commonly allow its subject a period of review and a further period to file any patent applications and such with the CRO; it is still common, however, to restrain any publication.

Termination Arrangements and Consequences

The grounds for termination of the agreement should be set forth clearly in the agreement. Typically, a clinical research agreement may be terminated in the following circumstances:

- At any time by the sponsor's written notice to the CRO;
- Where protocol-approval difficulties arise with the appropriate ethics committee or regulatory authority;
- In the event that the CRO becomes insolvent, bankrupt, or enters into liquidation;
- In the event of a material change of management or control of the CRO;
- If a party breaches any of the agreement's terms and fails to remedy that breach within a stated period; and/or
- In the event that subject recruitment fails to meet the requirements set forth in the contract.

Where the agreement is terminated at the sponsor's discretion, the question arises as to what compensation is payable to the CRO. Usually, the agreement provides that the CRO will be reimbursed for the cost of all work performed to the date of termination, along with any additional expenses for which the CRO has incurred commercial commitments that cannot be reasonably avoided (for example, if the CRO has also committed to purchasing supplies or equipment). The sponsor will not normally be liable for the CRO's lost profits, unless the CRO raised this issue specifically and negotiated to include such a provision in the contract. The sponsor (and the CRO) will, however, normally exclude liability for any such economic or indirect loss as discussed previously.

Other Legal Clauses

Commonly known as "boilerplating," an agreement with the CRO typically includes other terms of an administrative nature. These include the terms that follow.

Proper law and dispute resolution. The parties should stipulate which jurisdiction is to decide the dispute and which law will apply. Where the parties wish to go to arbitration or use alternative dispute-resolution procedures, these desires should be specified in detail.

Status of the parties. It is usual to provide that nothing in the contract research agreement shall constitute a partnership or agency, and that the CRO has no right or authority to enter into any contract or assume any other obligations on behalf of the sponsor.

Severability. If the parties intend to include any liability restrictions or exclusions, it is wise to include a severability provision in the agreement, stating that if a liability provision is determined to be illegal or unenforceable by any court, then the remaining provisions will be severable and enforceable as long as the agreement does not fail in its essential purpose.

Amendment. The agreement should provide that no modification, waiver, alteration, or amendment shall be valid, unless set forth in writing by both parties.

Waiver. It is a common feature of commercial life that one party may agree, perhaps reluctantly, to the request of the other, and promise that he will not insist upon performance according to the strict letter of the contract. This arrangement is usually referred to as a waiver. An agreement often requires a provision stating that any failure by a party to enforce or strictly require the other to observe and perform any terms of the agreement will not constitute a waiver or prejudice any other rights under the agreement.

Force majeure. It is usual to provide that neither party will be liable for any loss or damage resulting from its failure or delay in performing its obligations where that failure or delay arises from circumstances beyond the party's control. The party relying on force majeure should be required to serve written notice on the other party, providing a detailed estimation of the period that such prevention or delay will continue, and to minimize the disruption.

Entire agreement. In order to avoid the problems of misunderstandings arising from negotiations and the question of oral contracts between the parties, it is prudent to provide that the written contract signed by the parties supersedes and cancels all prior agreements, understandings, and negotiations in connection with it. This clause seeks to ensure that an oral variation to the contract is impossible, and to avoid a party bringing action against the other for precontract misrepresentation.

Notice. Where any notice by one party to the other is required, the agreement should specify how such notice should be served. If notice can be sent by e-mail or facsimile, it would be prudent to provide that a copy must be sent by post within twenty-four hours after the e-mail or facsimile has been transmitted.

CHANGE MANAGEMENT

The agreement should contain provisions allowing the parties to monitor and review the performance of the other party's contractual obligations.

The flow of information between the parties must also be addressed. The substance and frequency of meetings between the parties, along with an indication of who is expected to attend them, should be specified in the contract. The contract should further provide that minutes be kept to record the understandings between the parties during the course of the relationship.

A related issue is dispute resolution. The agreement should specify a hierarchy of management levels at which disputes should be discussed, as well as timelines for those discussions. There should be several levels of staff seniority through which the dispute can be escalated to prevent an overhasty rush to the courts.

Failing dispute resolution, the agreement could also require expert determination, mediation, or binding arbitration as an alternative to litigation. The agreement should be written to ensure that exhaustive efforts are provided to find a solution to the dispute, making litigation or arbitration the very last resort.

Contractual change management provisions should therefore cover at least the following elements:

- Frequent meetings between the sponsor and the CRO to review the service and address opportunities for improvements;
- Formal procedures for changing proposals and how they should be responded to;
- A set negotiation mechanism;
- Documenting and signing off on an agreed-upon change; and
- A requirement that the CRO should not unreasonably withhold its approval of a change.

The agreement should also ensure that the sponsor makes regular site inspections to relevant study centers and facilities.

EMPLOYMENT ISSUES

Of all the legal issues arising in an outsourcing arrangement, this is the most complicated, and specialist employment law advice should be taken. The following is only an overview and a "snapshot" based on the law at the time of writing. Specialist legal advice is imperative where there is any expectation that transfer of undertaking rules might apply. The application of the Transfer of Undertakings Protection of Employment (TUPE) Regulations of 1981 to an outsourcing arrangement is by no means certain. Further, there are areas of uncertainty in TUPE's effects if it does apply. For the purposes of this discussion, while UK legislation is referred to, these issues are applicable throughout the European Union as a result of the Acquired Rights Directive (77/187).

If TUPE does apply to any outsourcing, the sponsor's employees who are assigned to the outsourced business will transfer automatically by law to the CRO. Individuals who are engaged in the business but are not employees of the sponsor (such as self-employed or contract staff) will not be protected by the regulation.

Employees are transferred on their existing terms and conditions of service (except for pensions) and they retain the benefit of their previous continuous service. Any change to those terms and conditions for a transfer-connected reason will be ineffective, whether or not the change is beneficial to the employee and whether or not he or she consents. Often, a pragmatic approach is taken—if the employee's overall package is no less favorable, then he or she is unlikely to complain. As stated above, pensions are not covered by TUPE, but it is customary for a CRO to make comparable arrangements.

If an employee is dismissed in connection with the transfer, it will automatically be considered unfair dismissal unless the employee is dismissed for an economic, technical, or organizational (ETO) reason entailing changes in the workforce. The mere fact that the CRO wants to use its own staff is not an ETO reason.

It is a legal requirement that representatives of employees are informed and consulted on a transfer. It is generally best to explain early and openly that the outsourcing is a business and economic decision. Otherwise, the employees tend to find out for themselves and are occasionally misinformed, potentially causing motivation problems. The sponsor should do all it can to keep its employees informed and obtain the employees' consent to the arrangement it is trying to put in place.

TUPE applies to the transfer of an undertaking. The key question is whether there is a transfer of an economic entity (defined by the European Court of Justice as an organized grouping of persons and assets facilitating the exercise of an economic activity that pursues a specific objective) that retains its identity in the hands of the CRO. If there is, TUPE will apply.

It has generally been thought that first-generation outsourcing of services would necessarily involve a TUPE transfer. If the services can be contracted out, then they will usually be an economic entity. This economic entity would ordinarily retain its identity in the hands of the CRO. However, recent TUPE cases on second-generation outsourcing of services have held that there is no TUPE transfer unless assets are transferred or a significant part of the workforce is taken over at the same time. If the contracting out of services, such as contract research operations, does not include a sale or transfer of assets or workforce, then TUPE may not apply.

Second-generation outsourcing will not be a TUPE transfer unless assets or a significant part of the workforce is transferred from the outgoing to the incoming CRO. The law regards termination of the original contract and the signing of a new contract between the sponsor and the new CRO as a transfer between the outgoing and incoming CRO. If TUPE applies, the existing CRO avoids having to pay any redundancy costs. If, however, TUPE does not apply, then the existing CRO will be faced with redundancy costs.

TUPE may also apply when the outsourcing arrangement is terminated and the services are taken back in house.

In all cases, the sponsor will need to ensure that there are appropriate provisions in the contract to deal with the potential accrual of employment liabilities.

CONCLUSIONS AND THE FUTURE

This chapter is intended as a practical guide to important issues in negotiating and drafting a contract research agreement. While commercially these agreements involve an enormous degree of mutual trust, legally they are fairly well-tried and -tested arrangements.

However, we are also seeing the development of somewhat more innovative arrangements involving risk sharing, which goes beyond incentives and penalties for good or poor performance, although it is fair to say that conservatism and conflict-of-interest concerns are inhibiting the development of such arrangements (especially outside the United States). However, those we have seen tend to involve, to a greater or lesser extent, the licensing of intellectual property by the sponsor to the CRO, permanently or temporarily, possibly with an option for the sponsor to licence the candidate back, hopefully after it has progressed through the development stages. These agreements will thus involve the negotiation of terms associated with a licensing agreement, including:

- The IPRs involved and the scope of the license;
- Royalties, lump sums, and milestones;
- More detailed intellectual property warranties;
- Responsibilities for managing IPRs; and
- Ownership of IPRs on termination.

Another alternative is for CROs to take equity positions in, for instance, a smaller biotech company in return for waiving all or part of its normal fees.

The money involved in contract research agreements can be vast and it may be observed that the use of watertight legal contracts to protect a sponsor's investment is improving, albeit from a low base. Inevitably, U.S. examples have led the trend toward increasingly sophisticated agreements.

Richard Ashcroft

Ethical Considerations in Outsourced Clinical Trials

Scientific medicine has produced many great advances in human well-being and the relief of suffering. As it has developed, scientific medicine has grown dramatically in parallel with the growth of the pharmaceutical industry. One cause and consequence of this parallel growth is the development of medical research into a large industry in its own right—in both its academic and industry contexts. Medical research depends crucially on the participation of human subjects, both healthy volunteers and patients. Many authors have commented on the apparent tension between the clinical research coordinator's clinical obligations to the patient as a patient and his or her scientific obligations to the study, sponsor, scientific community, and future patients. We can also consider more widely the tensions between the interests of the scientific community, the business interests of the sponsor, the public policy interests of the state, the financial and professional interests of participating clinicians, and the short- and long-term interests of the host population. Resolving these tensions can become a complex matter of law, politics, and ethical debates on justice. Over all of these debates looms an awareness of the profound immorality of the human rights violations sometimes perpetrated by human experimenters.

These issues may seem remote from the concerns of the Site Management Organization (SMO), Contract Research Organization (CRO), or General Practice Research Network (GPRN), but they are relevant for three basic reasons. First, these are the issues that shaped the ethical guidelines and regulatory framework for clinical research over the past half century. Second, these principles and rules were drawn up before SMOs and CROs were thought of, or at least before they were widely used, so defining ethical principles for SMOs and CROs will depend, to begin with, on adapting these principles for SMO and CRO use. Third, because the SMO and CRO evolved as solutions to business process problems in the pharmaceutical industry, they are most easily theorized from a management science perspective. Nonetheless, they are, by their very nature, directly or indirectly involved in the medical care of patients. Hence, it is to medical ethics first and business ethics second that we must turn to make sense of SMO and CRO responsibilities in clinical research.

This chapter defines the basic ethical principles for SMOs and CROs and their use in clinical research. The next section reviews the main guidelines for research ethics as they have come to be defined in the past half century. The following sections review the underlying core principles of these guidelines and the values that influence their application. In the section entitled "Applications to Outsourced Clinical Research," I apply these values and principles to the specific problems of outsourced clinical trials. I review briefly why outsourcing has become important, and then consider the ethical issues that arise in outsourcing recruitment, trial management, and data and safety monitoring. The conclusion looks briefly at how research ethics committees will apply these principles.

ETHICAL GUIDELINES IN CLINICAL RESEARCH

There exist a remarkable number of guidelines on healthcare ethics. A recent survey, listing the guidelines merely by title and issuing institution, extends to twenty-one pages (Fluss 1999). However, for our purposes, there are a handful of key guidelines on the ethics of clinical research that define the core principles.

For historical reasons, the Nuremberg Code is still widely referred to. This was promulgated in 1947 as part of the judgments issued at the "Doctors' Trial" where a number of doctors and scientists were convicted of crimes against humanity because of some inhumane experiments carried out on inmates of the Nazi death camps. The Code summarized the reasons for holding that these experiments were unconscionable. The key principle in the Code was the principle of voluntary, uncoerced consent to participation in the experiment (for a brief history, see Moreno 1999, chap. 3; for the code itself see Brody 1998, app. 1).

The Nuremberg Code, while powerful and influential, is not the most important document in practice. Because the Code insists on consent, it rules out substantial valuable experimentation on people unable to consent (for instance, young children and people who are unconscious or otherwise mentally incompetent to consent to the research). For this reason, in 1964, the World Medical Association issued the first edition of the Declaration of Helsinki, which is more detailed and wider in scope than the Nuremberg Code (World Medical Association 2000). This document has undergone a number of revisions. The most recent of these to be accepted was in 2000, and is being debated as this book goes to press. (See Schüklenk and Ashcroft 2000 for a commentary on this debate.) The Declaration originally concentrated on therapeutic medical research and nontherapeutic biomedical research (chiefly physiology and pharmacology); today it still focuses on biomedical research, but the range of types of biomedical research covered is much broader, including genetics, psychiatry, human reproduction, epidemiology, and even health services research. Similarly, research is now possible, within strict limits, on young children, the learning disabled, the mentally ill, and other classes of vulnerable subjects. Moreover, the site of clinical research is no longer only the laboratory, clinic, or consulting room, but may include the community or research conducted in several international centers. Research in the developing world is increasingly controversial and poses great challenges to ethicists, sponsors, and clinicians (Nicholson 1999; Nicholson and Crawley 1999).

The Helsinki Declaration is the main statement on principles of ethical clinical research, although other documents of various kinds (for instance, the Council of Europe's Convention on Biomedical Ethics, the U.S. Food and Drug Administration's regulations, the International Committee on Harmonization's Good Clinical Practice framework, and the Council of the International Organizations of Medical Sciences guidelines). Furthermore, the clinical research agenda is largely determined by national and international law and regulatory policy. Nonetheless, the core principles of ethical clinical research are few and simply stated (Smith 1999).

ETHICAL PRINCIPLES

The chief principle for research ethics since the Nuremberg Code is the informed, voluntary consent of the participant. This requires that participants know what they are being asked to do, what the likely consequences and possible risks and benefits of the research treatment are, and what their alternatives are. It also requires that they be able to make a choice, both in the sense of being free from coercion or improper inducement and in the sense of possessing the capacity to consent. Capacity to consent requires the ability to understand, retain, and reflect on the information necessary to make a reasonable decision about the choice facing them. It does not, by the way, require that a participant's decision actually be reasonable. These conditions imply that participants need to know that they are taking part in research, and that they may need to undergo additional tests as part of the research that they would not need ordinarily, and that their medical records and perhaps other information may be used by persons other than those involved directly in their clinical care.

Consent is both an ethical and a legal concept, and the precise requirements of the law and professional bodies vary from place to place and from time to time. Here, it is worth noting that this ethical principle is simple and, while research ethics guidelines have developed, the centrality of informed consent has remained a constant.

The second main ethical principle of research ethics is proportionality of risk and benefit. No one uses this phrase in a mathematical sense strictly; it is intended to be understood metaphorically. What the phrase means is that the risks posed by the research be reasonable, given the expected benefits to the participant. So, for example, a patient with an advanced and inoperable cancer may be offered participation in a Phase I or II clinical trial that offers some possibility for tumor remission at the price of great risks of side effects and discomfort. The patient may decide that the risks are reasonable, given the possibilities of the treatment if it is effective. Some people think such trials would be unjustified but for their altruistic aspects, since the probability of survival for such patients is slim, but that is beside the point. There is a need for such volunteers in early phase trials of anticancer drugs precisely because the risks to a healthy volunteer would not be reasonable. While it is an interesting philosophical point to consider whether a sufficiently high payment or other inducement would compensate some healthy volunteers, in practice, ethics committees would not regard financial compensation as sufficient to permit this kind of risk-bearing by volunteers. This decision has to do with worries about exploitation rather than the possibility of trading cash for risk as such. Exploitation worries can be considered as a social justice issue or as an issue about induced or coerced—hence nonvoluntary—consent (Wertheimer 1996).

These two ethical principles (consent and risk-to-benefit proportionality) pull together almost everything written about research ethics until recently. To these principles should be added a third, which may or may not be subsumed under the first two: Clinical research should be fair to its participants and to the community, society, or country hosting the research. This is a principle of justice in research (Kahn, Mastroianni, and Sugarman 1998). This principle can be difficult to interpret, and its interpretation is hotly contested; a good recent example is the debate over whether or not researchers and sponsors are obligated to provide AZT to third world HIV vaccine trial participants who become infected as a result of participating in the trial. The fact is that AZT is not normally available to these participants by reason of its expense (Schüklenk and Ashcroft 2000).

Finally, the research itself should be well-designed, likely to elicit a definite result, and likely to answer a question to which we do not already know the answer. This may not look like a principle of ethics, but its intention is to prevent the unnecessary exposure of people to the risks of research and the unnecessary exclusion of the control group from treatment already proven effective. "Unnecessary" here means either unnecessary because already done or unnecessary because useless or incompetent. From this principle flow some other controversies: Should there be an obligation on clinical researchers to publish or share their results to prevent other scientists from needing to reinvent the wheel (Horton and Smith 1999)? Should clinical researchers prove, by preparing a systematic review, that the research question has not already been answered (Savulescu, Chalmers, and Blunt 1996)? More broadly, is research done to satisfy regulatory requirements to be governed by the standards of the scientific community or by those of commercial and regulatory interests (Royal College of Physicians 1986)?

These issues are complex and are at the heart of the ethics of SMOs. We will return to them below. But to make sense of these principles and, better, to understand both their scope and how to apply them in practice, it is useful to step back for a moment and identify the fundamental ethical values they express.

ETHICAL VALUES IN RESEARCH

The ethical values of healthcare research can be grouped into three kinds: *patient-centered* values, *public interest* values, and *enterprise* values.

Patient-Centered Values

Patient- or participant-centered values relate to the fundamental importance of respect for persons. These values include the value of *patient autonomy* (the right to make one's own decisions, to make one's own judgment about what is in one's best interests, and to do so free from coercion, undue pressure, or unreasonable constraint) and the *fidu-*

ciary values of beneficence and nonmaleficence. The latter values underpin the obligations of the healthcare worker to act in the patient's best interests, subject to the
patient's informed consent.

It is important to note two consequences deriving from the values of autonomy and
of professional good faith (fiduciary values) (Rodwin 1993). First, they imply that
patients are to be respected as persons, rather than treated as business resources merely,
usefully employed in one's own commercial or academic projects. This principle is in
some tension with the SMO/CRO position as a patient recruiter and trial manager on
behalf of the sponsor: The SMO/CRO is under pressure to recruit patients in accordance with budget and time constraints that may create difficulties in the recruiter-
patient relationship, and the SMO/CRO may be competing against other suppliers
that have more access to patients or that are less scrupulous.

The second consequence of the autonomy and fiduciary values for SMO/CRO
work is that it is a mistake to consider patients as consumers of research or healthcare
services. There is no sense in which patients are free contractors, subject to the buyer
beware principle; moreover, healthcare workers are expected to act in their patients'
interests as well as their own. This is not to say that contractors cannot act in their own
commercial interests, nor that patients are all passive quasi-victims who need looking
after—quite the opposite can be true! Rather, it is to suggest that the doctor-patient
relationship is a professional-client relationship, not a consumer-supplier relationship.
This is so even in nontherapeutic or healthy volunteer research, since such participants
have a moral right, as a condition of their participation, to medical care for any symptoms or problems arising as a result of their participation.

Public Interest Values

The *public interest* values in research relate to the acknowledged public interest in the
existence of quality research and development in the fields of health, health care, and
health technologies. These values relate to the values of *justice, public good*, and *obligations to future generations.*

Justice requires nondiscrimination in recruitment and fair distribution of research
risks and benefits amongst the population in any society (Ashcroft 1997). The *public
good* requires that research be well-designed to answer a genuine and meaningful
research question. It can also be used as a limited justification for certain kinds of
research without consent. For example, in some circumstances, epidemiological
research can require access to patient records, or to a compilation of statistical information, in circumstances where patient consent would be difficult or impossible to
obtain for practical reasons. If the research is of sufficient importance, and the interests of the participants are protected in other ways (for example, by making patient
information anonymous), the use of this information can be justified.

Research interventions on patients who cannot consent by reason of incapacity can
sometimes be justified if it is in their best interests (roughly speaking, that the non-
research interventions would be unsuccessful, less beneficial, or are simply of unproven
and doubtful benefit in comparison with the research intervention). However, this is
a patient-centered justification. More controversial are those justifications of certain
kinds of psychiatric research trials in which the treatment is aimed not at the patient's
benefit, but at public safety. Other justifiable trials represent minimal risk to the

incapable subject and might benefit other patients with the same or similar conditions in the future. This is not a public interest justification per se, but an appeal to *obligations to future generations*. In some cases, the risk need not be minimal, although justifying a high-risk trial through an appeal to future good is highly controversial in law and in ethics. Good examples would be early-phase pediatric oncology trials in very young, terminal-phase patients. Such trials are almost never planned with an eye toward participant benefit, as this is very unlikely; rather, the hope is that they will benefit future patients in the same disease and age group.

Public interest and future generations considerations are very hard to adjudicate, and while there are guidelines produced by organizations such as the medical Royal Colleges and the Medical Research Council in the UK, there is no clear-cut international consensus, and the judgments that must be made normally fall to the participants or their relatives themselves and to research ethics committees. It is probably fair to say that, more often than not, the public interest in protecting research participants is better reflected by strict adherence to the patient-centered values than to attempts to interpret what "public interest" means in the research context.

Enterprise Values

The *enterprise* values in medical research are not widely discussed outside the pharmaceutical industry, although they deserve serious consideration, provided that they are seen as balancing the patient-centered and public interest values. Most commentators would agree that, in the absence of pharmaceutical industry nationalization, and given the various ways in which the industry is regulated through central price-fixing, medicinal products safety controls, and so on, there is a strong public interest in allowing the industry to carry out its research and development programs with relatively little further interference. Consequently, it is unreasonable to discriminate against industry research merely because it is commercial, or to overregulate its activities, or to regulate its activities in unclear ways. For instance, it is conceivable that we would like to prevent the exploitation of research subjects, but if we cannot frame clear and equitable regulations that define exploitation, then it would perhaps be better not to regulate at all.

At root, the enterprise values represent the legitimate interests of private individuals and corporations to carry out their lawful business without undue interference from the state. This may sound like advocating a free hand for the pharmaceutical industry, which is quite powerful enough as it is! In fact, the enterprise values also relate to a concern sometimes expressed by patients as well as researchers: that some opportunities to take part in research or gain access to experimental treatments are unduly limited by apparently ethical considerations (Schüklenk 1998).

APPLICATIONS TO OUTSOURCED CLINICAL RESEARCH

Why Outsource?

The idea of outsourcing in clinical research is, from a business point of view, very simple. All pharmaceutical companies structured along the traditional corporate model have divisions devoted to research and development, production, marketing, and corporate functions (personnel, finance, and so on). Science-led industries (like pharmaceuticals) often have very complex research and development (R&D) divisions. Drug development has many phases, from rather pure biochemical research into new chemical entities (NCEs) and drug design, through human pharmacology and pharmacogenetics, to regulatory affairs, late-phase clinical trials, and so on. Traditionally, pharmaceutical firms sought to grow through vertical integration (taking over biotechnology start-up firms or drug manufacturing plants, or through investment in private health care) and through horizontal integration (takeovers of rivals to ensure access to new product lines) (Towse 1994). However, these traditional strategies are increasingly seen as risky and uncertain in all industries, and pharmaceuticals has been no exception. Some reasons why the traditional strategy has become risky include poor grafting of diverse corporate cultures, incomplete understanding of the business of different technologies (is biotechnology really similar in business to pharmaceuticals?), difficulties in adapting to new markets, and so on (Ashcroft 1999).

For these reasons, pharmaceutical corporations increasingly find it worthwhile to use third party companies for particular activities where insourcing may be inefficient or ineffective. The financial press has paid considerable attention to this trend in recent months, but most of this has been focused on the relationship between pharmaceutical companies and high-technology firms in genomics, biotechnology, and molecular design (FT Life sciences—pharmaceuticals 1999; FT Life sciences—outsourcing 1999; Beyond the behemoths 1998). These relationships can and do involve ethical issues, but they are often remote from day-to-day patient care, or even from traditional healthy volunteers research (Ashcroft, Capps, and Huxtable 2000; Rothstein 1997; Weir 1998). In general, these issues fall into three groups: generic bioethics issues concerning the environmental safety of biotechnology; public policy issues concerning the ethics of drug research strategy and long-term health service planning (priority setting is a justice issue); and issues concerning the acquisition and maintenance of tissue and DNA banks. For the first and third of these, see Murray and Mehlman 2000; there is little on priority setting in research, but see Dresser 1999; Robert, Stevens, and Gabbay 1999; and Mowatt et al. 1997.

Outsourcing of clinical trials can meet various needs in the rationalizing of pharmaceutical industry structure and the structure of particular pharmaceutical companies. Of special importance for us here, companies are increasingly interested in outsourcing patient recruitment, clinical trial management, and data and safety monitoring. These introduce no new ethical issues, since most that arise in the new industry structure were already implicit and problematic hitherto. However, they do introduce old ethical issues in new and sometimes more complex ways.

Outsourcing Recruitment

Traditionally, industry trials recruited research subjects in two ways. Healthy volunteer studies were very often done in house by pharmaceutical physicians, who were bound by the ethics codes of their professional organizations and, increasingly, independent ethics committees. The latter is now obligatory under Good Clinical Practice (GCP) regulations in any case. Many healthy volunteer studies and most clinical trials were done through industry sponsorship of research carried out in healthcare or university settings (hospitals or general practice clinics, for instance). In the former case, recruitment was done by advertising, and in the latter, recruitment was done through contact between the patient and his or her physician, with the clinical trial positioned as part of the patient's treatment plan. The physician acted under medical obligations to the patient, while having a financial and trial management relationship with the sponsor, and very often a further financial and management relationship with his or her employer, so that research contracts were managed by the host institution. In theory, these arrangements protected patients and avoided conflicts of interest; in practice, this was not always the case, and there has sometimes been suspicion that keen researcher/physicians have neglected their main duty to their patients on occasion. Alongside this worry, there is anecdotal evidence in the medical press to suggest that the main type of research misconduct relates not to the unethical treatment of patients, but rather to fraudulent practice (fudging or inventing research data, fictitious enrollment of patients, and so on) (Lock and Wells 1996).

Outsourcing recruitment is subject to similar pressures, made more acute. Research hospital or university physicians normally have other sources of funding than their clinical trials work, whereas SMOs and CROs that recruit patients depend entirely on their ability to do so and to sell this service to industry. Moreover, if they operate as managers of recruitment, offering access to networks of freelance physicians and research nurses rather than recruiting and providing the treatment service directly, then there is a question about whether they are bound by canons of medical ethics at all. If, on the other hand, they provide physicians with research nurses or if they offer clinical trial treatments at all, then there needs to be some mechanism for ensuring that these professionals are encouraged to consider their patients only from the point of view of their suitability for the trial and that these professionals are protected from pressure to treat patients as resources.

Another source of recruitment difficulty involves building databases of patient information for trial recruitment purposes. According to the traditional pattern, patients were recruited by a physician they either had a preexisting clinical relationship with or had been referred to by another physician with whom they had a clinical relationship. In either case, their personal information was transferred in accordance with medical confidentiality, security of records, and fair processing as defined by data protection legislation. Current best practice specifies that a patient should be recruited by his or her general practitioner or by the specialist he or she normally consults for treatment.

There is legal and ethical uncertainty in this area, but research ethics committees have tended to argue that constructing databases of patient information that are held for purposes other than ongoing patient care is questionable, and needs special justification. Databases held by SMOs that are used as a source of patients for direct approaches, thereby circumventing the normal clinical relationship, and which are treated as an informational resource in commerce cannot normally be justified. This is not only an issue for commercial research; academic and government epidemiologists are increasingly finding some kinds of research difficult for data privacy and confidentiality reasons. Research ethics committees refer to principles of consent (has the patient agreed implicitly or explicitly to this use of data?), reasonableness, privacy, nonmaleficence (unexpected or nonmedical approaches can be distressing), and absence of fiduciary obligation (an SMO owes no medical duties of care, confidentiality, or disclosure of medical records to patients and so its legal obligations are very unclear). The bottom line here is that SMOs and CROs are often perceived by research ethics committees as farmers of patients merely, who are interested in commercial questions first and last, and in medical questions as a means to commercial ends only. This perception is in almost all cases very one-sided, ignoring all the improvements in patient care an SMO may offer, but it needs to be reckoned with nonetheless.

Finally, related to recruitment and records management, much recent debate has focused on the principles covering obtaining and maintaining human tissue and genetic material for banking, especially where these samples can be developed into a marketable product or where the samples can be linked back to patients or their records. The legal and ethical position here is unclear, but the UK Medical Research Council has published valuable guidelines on human tissues and on personal information (Medical Research Council 2000; 2001).

Clinical Trial Management

The principle that outsourcing clinical trial management provides a way of improving the quality assurance (QA) of clinical trials is sound and ethically praiseworthy. However, there are risks. As with recruitment, the commercial imperatives to meet sponsor deadlines and standards can prompt corner cutting and poor patient care. In particular, standards of informed consent may fall. For trials that take place far from the sponsor's main office, local conditions or local agent activities may fall from the documented standard—this has been questioned in some multinational studies, for example. To the extent that the introduction of an SMO as a third party into the sponsor-researcher relationship or as a fourth party into the sponsor-researcher-patient relationship amounts to the introduction of an additional management tier, the normal risks of additional bureaucracy apply. A recent similar instance is the introduction of multicenter research ethics committees in the UK, which have tended to extend the time to approval, rather than reduce it (Alberti 2000).

An issue that periodically causes controversy is whether researchers or sponsors are obliged to continue supplying experimental treatments after the end of the trial, even if the trial is terminated early (for business or statistical reasons) or even if the trial demonstrates no benefit. Like issues in clinical trial design (fixing of endpoints or surrogate endpoints, determining the number of trial arms, the kind of controls used, and so forth), the decision whether or not to continue supplying experimental treatments would normally be fixed by the sponsor, or, occasionally, by an academic partner or the supporting health service. However, it is now possible to outsource this element of trial management. Research ethics committees are reluctant to approve trials of "me too" drugs or trials suspected of being done for marketing or regulatory purposes merely (rather to answer a genuine research question) (Savulescu, Chalmers, and Blunt 1996). Some commentators have questioned whether this is a reasonable attitude (Bosanquet 1999; Kettler 1998). Nevertheless, the question of whether it is fair to provide someone suffering from a chronic illness or disability with treatment that is found to benefit them and then to withdraw it in order to force the hand of the purchaser is a serious one, and one that is hard to answer (admittedly, the ethical fault may lie with the purchaser, too).

Current opinion holds that this question of posttrial treatment should be discussed with the patient prior to entry. The issue for the outsourcing agent is that, while the sponsor may feel an obligation to a physician and his or her patient, this may well not be the case when the sponsor deals with an SMO, and the SMO deals with the physician.

Data Monitoring and Safety

The use of CROs to handle trial data management and product safety monitoring is now well established. There are three important ethical issues here: stopping trials early, managing unfavorable results, and data ownership.

Occasionally, sponsors will withdraw support from trials for business reasons—for example, recruitment is too slow or the market too small (Ashcroft 1999). More often, trials may be terminated for scientific or technical reasons (problems with design or statistical power), or because interim evidence suggests that the treatment is either very much more effective than the control treatment, or very much less effective, or has unexpected, unexpectedly serious, or frequent side effects (Baum, Houghton, and Abrams 1994; Pocock 1993). It is now regarded as essential good clinical practice to decide in advance what clinical cessation rules should be applied in a clinical trial, and for the trial data to be analyzed and reviewed on a regular basis by an independent data and safety monitoring board (DSMB). These rules embody ethical judgments about just how much more effective or ineffective the trial treatment needs to be or how serious or frequent the side effects need to be for trial to be suspended. It is generally agreed that stopping a trial early because of inefficacy or serious side effects should be easier and take place more quickly than stopping a trial early because of greatly superior effectiveness. In addition, if greatly superior effectiveness is expected, then this can be incorporated into the power calculation for the trial, and smaller numbers of patients are needed. However, DSMBs are not fail-safe; they can be misled by random error, thus preventing access to effective treatment or promoting access to ineffective treatment.

These issues are obviously ethical and would arise in any clinical trial. In some respects, it is easier for an independent contractor to make these judgments, since they are not subject to the good money after bad fallacy—if a company has been involved in developing a drug for a long time, they may be unwilling to give up on it, even when they should.

Although independent contractors might face fewer pressures in some areas, they probably face more in others. The CRO can face tremendous pressure from the sponsor to suppress certain kinds of information arising from the contracted analysis; for instance, where a treatment is found to be efficacious but not significantly more so than a competitor treatment, and other trial evidence suggests a more favorable assessment of the drug. Publication bias is another source of pressure on CROs to suppress data produced in trials; this is a big problem in both industry trials and academic research (Chalmers 1990).

CROs and SMOs could learn from the statistical profession here. For example, the Royal Statistical Society (RSS) Code of Conduct (1993) says, at Title 6, that members of the society (fellows) "should not allow any misleading summary of data to be issued in their name." Further, at Title 3, the Code specifies that "fellows shall carry out work with due care and diligence in accordance with the requirements of the employer or client *and shall, if their professional judgement is over-ruled, indicate the likely consequences"* (emphasis added). The suggestion is that CROs should set out in their standard contract what their terms of business and professional obligations are and reserve the right to withdraw their imprimatur from the report if they feel that report is being used misleadingly. However, the RSS also indicates that responsibility for further misconduct will lie with the sponsor, as is implicit in Title 3 (cited). Title 16 (salaried employment) of the Code says, "if the conflict cannot be resolved satisfactorily the public interest and professional standards must be paramount," and Title 17 (private practice) clearly states that "fellows . . . have the right of disengagement in the face of a dilemma involving professional standards or conscience."

The third ethical issue in data monitoring is ownership of data. We have already touched on the issue of access to patients' medical records and personal information. It should be clear that medical records regarding the medical condition and treatment of participants are legally open for the patient to see and copy (for instance, in the interests of apprising their doctor of their medical condition, or if they are pursuing legal action). Nonetheless, the *ownership* of these records is legally complex. Typically, CROs are not concerned with patients' personal information; rather they are more concerned with controlling aggregate data, which is unlinked from patients and rendered anonymous, relating to statistical data from the trial. Clearly, this is a contractual matter between the researcher, the sponsor, and the CRO and/or SMO involved. Nonetheless, such contracts must consider the ethical issue of publication bias discussed above (including the publication of regulatory affairs documents). There are legally and morally complex issues of intellectual property here, which are too large to consider adequately in this chapter.

CONCLUSION

This chapter has reviewed the main ethical issues involved in using or operating an outsourcing organization in clinical trials. Since this field is developing, this chapter is unlikely to be totally comprehensive or definitive! Nonetheless, it is clear that outsourcing is here to stay, and since it can contribute greatly to improving the quality of drug research and development, it should be applauded. While the ethical problems surveyed here are sometimes difficult to resolve, sensible use of the existing ethics and Good Clinical Practice guidelines, together with a partnership with the research ethics committees, should ensure that most problems will be manageable.

Research ethics committees have up to now been suspicious of SMOs and have had little contact with CROs. They tend to focus on the patient-centered values within commercial research and tend to be dubious about enterprise values, although academic researchers occasionally make successful appeals to public interest values. This situation can change if SMOs and CROs are better prepared to inform research ethics committees about what they do and how they do it, and about what kinds of contractual protection measures are in place to protect the patients' interests. In general, if a trial is well-designed and well-managed, and if patients are happy, the research ethics committee will discover that the SMO is possibly a safer pair of hands, ethically speaking, than the freelance researcher.

REFERENCES

Alberti, K. G. M. M. 2000. Multicentre research ethics committees: Has the cure been worse than the disease? *British Medical Journal* 320: 1157–1158.

Ashcroft, R. E. 1997. Human research subjects. In *Encyclopedia of applied ethics*, ed. R. Chadwick, 627–639. 2 vols. San Diego: Academic Press.

Ashcroft, R. E. 1999. Stopping clinical trials early. *Clinical Research Focus* 10(8): 36–39.

Ashcroft, R. E., B. J. Capps, and R. Huxtable. 2000. The advisory and regulatory framework for biotechnology in the UK. In *Encyclopedia of ethical, legal and policy issues in biotechnology*, eds. T. H. Murray and M. J. Mehlman, 747–761. New York: John Wiley.

Baum, M., J. Houghton, and K. Abrams. 1994. Early stopping rules: Clinical perspectives and ethical considerations. *Statistics in Medicine* 13: 1459–1469.

Beyond the behemoths: Mergers and acquisitions in the pharmaceutical industry. 1998. *The Economist*, 21 February.

Bosanquet, N. 1999. European pharmaceuticals 1993–1998: The new disease of innovation phobia. *European Business Journal* 11: 130–138.

Brody, B. A. 1998. *The ethics of biomedical research: An international perspective.* New York: Oxford University Press.

Chalmers, I. 1990. Underreporting research is scientific misconduct. *Journal of the American Medical Association* 263: 1405–1408.

Dresser, R. 1999. Public advocacy and allocation of federal funds for biomedical research. *Milbank Quarterly* 77(2): 257–274.

Fluss, S. S. 1999. International guidelines on bioethics: Informal listing of selected international codes, declarations, guidelines, etc. on medical ethics/bioethics/health care ethics/human rights aspects of health. *The EFGCP News*, December (supplement, salve 2).

FT Life sciences surveys on pharmaceuticals and biotechnology industries. 1999. *Financial Times*, 31 March.

FT Life sciences survey on outsourcing in the pharmaceutical industry. 1999. *Financial Times*, 15 July.

Horton, R. and R. Smith. 1999. Time to register randomised trials. *British Medical Journal* 319: 865–866.

Kahn, J. P., A. C. Mastroianni, and J. Sugarman, eds. 1998. *Beyond consent: Seeking justice in research.* New York: Oxford University Press.

Kettler, H. E. 1998. *Competition through innovation, innovation through competition.* London: Office of Health Economics.

Lock, S. and S. Wells, eds. 1996. *Fraud and misconduct in medical research.* London: BMJ Publishing.

Medical Research Council. 2000. *Personal information in medical research.* London: Medical Research Council.

Medical Research Council. 2001. *Human tissue and biological samples for use in research: Operational and ethical guidelines.* London: Medical Research Council.

Moreno, J. D. 1999. *Undue risk: Secret state experiments on humans.* New York: W. H. Freeman and Co.

Mowatt, G., et al. 1997. When and how to assess fast changing medical technologies: A comparative study of medical applications of four generic technologies. *Health Technology Assessment* 1(14): 1–150.

Murray, T. H. and M. J. Mehlman, eds. 2000. *Encyclopedia of ethical, legal and policy issues in biotechnology.* New York: John Wiley.

Nicholson, R. H., ed. 1999. Revising the Declaration of Helsinki. *Bulletin of Medical Ethics* (August) 150 (special issue).

Nicholson, R. H. and F. P. Crawley. 1999. Revising the Declaration of Helsinki: A fresh start. Meeting report. *Bulletin of Medical Ethics* (October) 151: 13–17.

Pocock, S. J. 1993. Statistical and ethical issues in monitoring clinical trials. *Statistics in Medicine* 12: 1459–1475.

Robert, G., A. Stevens, and J. Gabbay. 1999. "Early warning systems" for identifying new healthcare technologies. *Health Technology Assessment* 3(13): 1–100.

Rodwin, M. A. 1993. *Medicine, money and morals: Physicians' conflicts of interest.* New York: Oxford University Press.

Rothstein, M. A., ed. 1997. *Genetic secrets: Protecting privacy and confidentiality in the genetic era.* New Haven: Yale University Press.

Royal College of Physicians. 1986. *The relationship between physicians and the pharmaceutical industry.* London: Royal College of Physicians.

Royal Statistical Society. 1993. Code of conduct. London: Royal Statistical Society.

Savulescu, J., I. Chalmers, and J. Blunt. 1996. Are research ethics committees behaving unethically? Some suggestions for improving performance and accountability. *British Medical Journal* 313: 1390–1393.

Schüklenk, U. 1998. *Access to experimental drugs in terminal illness.* Binghamton, NY: Pharmaceutical Products Press.

Schüklenk, U. and R. E. Ashcroft. 2000. International research ethics. *Bioethics* 14(2): 158–172.

Smith, T. 1999. *Ethics in medical research: A handbook of good practice.* Cambridge: Cambridge University Press.

Towse, A., ed. 1994. *Industrial policy and the pharmaceutical industry.* London: Office of Health Economics.

Weir, R. F., ed. 1998. *Stored tissue samples: Ethical, legal, and public policy implications.* Iowa City: University of Iowa Press.

Wertheimer, A. 1996. *Exploitation.* Princeton, NJ: Princeton University Press.

World Medical Association. 2000. Declaration of Helsinki: Ethical principles for medical research involving human subjects. http://www.wma.net/e/policy/17c.pdf.

Lucien Steru

Avoiding and Managing Conflict in Outsourcing Clinical Research

The purpose of this chapter is not to tell the reader how to avoid disasters in clinical research. Its objective is, rather, to give advice on how to avoid, prevent, control, and even detect those areas of human relational conflicts that can have such negative effects on clinical research.

In the past decades, a conflict arising in a clinical program was generally considered to be an unpleasant incident. Today, it is rapidly identified as an economic threat. The consolidation that occurred in the nineties, both in the pharmaceutical industry and in the contract research organization (CRO) industry, created both larger players and more tense employees because of the inherent pressures it brings. The pharmaceutical industry reacts to health cost containment by increasing pressure to accelerate drug development. Also, the attitude of the medical profession regarding clinical research, which is often now a significant source of financing, allows a more business-oriented triangular cooperation between sponsors, CROs, and clinical investigators.

It is important to remember three specific features of clinical research: (a) However high the stake of a particular program may be, it still involves patients' health, billions of dollars, and the making or breaking of careers; (b) the very core of clinical research, inclusion and patient assessment, is not under the direct control of either the sponsor or the contractor, being mediated instead by investigators interacting with patients; and (c) clinical research lasts generally for years. These factors combine to create an environment of high anxiety, leaving humans feeling vulnerable.

We have to recognize what is seldom emphasized, because it can be disturbing that, however high the scientific, regulatory, and business stakes are, clinical research, and especially contract clinical research, still is mainly a human venture.

It is based on the personalities, attitudes, and cultural factors of investigators (and their patients), sponsor representatives, and contractors, and more and more frequently, it crosses borders and languages. The higher the stakes, the higher the stress and the more explosive the atmosphere, thus preparing the ground for conflict. What follows may look very unspecific and very basic. Yet controlling the factors highlighted is often as difficult as avoiding conflict within a marriage.

We will define the conflict as we understand it, and look at its dynamics, describe the background factors facilitating its occurrence, describe some of the main symptoms that herald the conflict, and give some tips to defuse it. Finally, we shall suggest ten corporate preventive actions and attitudes to avoid conflicts, or to solve them before they get out of control.

THE CONFLICT

What Is a Conflict?

Our operational definition is that a conflict consists of the hostile feelings generated by the pejorative perception of a problem, its causes, and its (mis)handling. It could be viewed as a disease with an objective problem becoming infected, the severity of which can be variable, and which creates negative feelings.

In most of the cases (but not all), the conflict is triggered by the sponsor. It starts with an identified *problem*. It can be a central problem: Patient or investigator inclusions are slow and no credible plan is presented on how to catch up. It can be a collateral problem: A progress report has been sent with another sponsor's name on it. Not that it matters much, but this begs the question "Is everything out of control?" Or the source of conflict can be a simple dislike or mistrust issue.

What is important is that the problem's nature will usually not evolve for a while, i.e., it will not receive immediate attention/treatment, thus validating the hostile feeling. In other words, we would say that the *infection* of the problem will generate the conflict: The attitudes of one party change to hostility, provoking the *contagion* on the other parties. That can remain "local," say at a clinical research associate (CRA) level, or turn into a "sepsis" type of syndrome, infecting investigators, upper management, and so on. And nothing will tell you how badly it may evolve. Add to a minor scratch some poisonous remarks ending with "they are ruining our study," involve upper management, calling on them to save the program, and see how far infection can go.

The "infection" analogy compares the initial "problem" to the irritating cause, leading to the "conflict" as the problem gets infected by the hostile feelings. Then contagion will generalize the hostile climate and possibly induce offensive actions (withholding payment, delaying data transfer), jeopardizing the project forever, and spreading within and across the sponsor and CRO organizations.

Let us accept in these pages that whatever happens, a conflict has to be avoided to minimize the damage. Even if a sponsor discovers that it had hired an inadequate CRO, if both sponsor and CRO are familiar with service management, they could go their separate ways with a win-win attitude, "win" meaning limiting the damage on both sides.

What Is Not a Conflict?

We insist that a problem should not necessarily lead to a conflict, even if it is a serious problem, such as a statement that "our selection criteria are such that we don't have patients." The good news is that when you hear "we" covering sponsor and CRO, this means we are together on the same side of the problems, fighting them hand in hand. Similarly, a cold relationship does not necessarily generate a conflict, although it is often the source of one.

A contractual disagreement is not a conflict either. As long as both parties negotiate in good faith, there is hope of an amicable solution. But poorly thought or badly drafted contracts may cause irritation, which eventually can turn into a conflict.

We could add further examples, but the psychopathology is clear: As long as hostile feelings are not disrupting the sponsor-CRO relationship, there is no conflict. But if things evolve "naturally," conflict will eventually burst, due to the lack of appropriate care.

Psychodynamics of a Conflict

In a situation of conflict, confidence (generally of the sponsor's representative in the CRO's personnel) is either absent or has been betrayed. Confidence is the antidote to hostility. For if you trust me, I know I (we) will find a reasonable way to solve the issue. We may have done it before. But if you don't trust me, or if you have lost confidence, then you will expect the worse, and act in consequence even if I have not had the appropriate chance to respond (since you anticipate I will not).

Hostile feelings in the sponsor-CRO relationship are often based on the sponsor's wish to *punish* the CRO as a whole (a feeling that "they won't get away with it and with our money," for instance). There is a positive side to this hostility, namely, the implicit sponsor feeling that "I've got to save this project against the CRO's (and maybe my own firm's) incompetence, dishonesty, and so on." When CRO personnel are hit by the hostility, the response is equally irrational; there is wounded pride ("after all I did for them") and desire to retaliate. So even if the problem was rational, the behavior generally stops being rational. As the ancient Romans put it: "Anger is a short insanity."

The fascinating phenomenon is that these irrational feelings are seldom visible; they are hidden by corporate and social etiquette. But after a while, in the bad cases, both corporations are dealing with what looks like a major dispute. You know you have reached the worst case when legal departments get involved.

Let us see how to avoid this kind of dispute and prevent the ugly exchange of threats: "you won't get your money" versus "you won't get your data."

First, there are two things to keep in mind that may help avoid the natural response that escalates the situation.

If it is not rational, it is not serious

When some noise alerts the CRO management, the first-line personnel tend to cover up: "We're not doing so badly; they have nothing serious against us." The use of "they" to refer to the sponsor is a sure sign of a problem.

The author's company used a management consultant to train its staff in CRO/customer relations. He linked the occurrence of crises to that of the collecting of the

"Green Shield" or loyalty stamps by customers in shops or gas stations. With a full page of stamps, the customer receives a free gift. If we reverse the meaning of the stamp (not a reward here, but a small frustration) and the gift (here, a crisis), we can use the analogy: The CRA (or a project manager) may, after an untoward incident with a staff member from the CRO (or sponsor), stick a stamp in that staff member's page, but otherwise not fight the person perceived as the troublemaker. Perhaps if the incident is serious, three stamps are given. The relationship then continues, apparently without hostility or conflict. However, eventually, as soon as the page is full, this CRA overreacts, not to the specific next incident but to the pent-up frustration contained within the whole page.

The staff member cannot understand the irrationality of it all, thinking "I didn't deserve this; it was just a minor incident." Thus, an opportunity to defuse the conflict is lost.

Symmetric escalation

This term describes a natural behavior, but one that can be very counterproductive. Anybody who has spoken in front of an undisciplined audience can understand the following example. The audience does not listen, whispering amongst themselves. Thus, the speaker speaks louder because of the background noise. The background noise itself leads the audience to whisper louder, then to talk, and eventually to shout. The speaker himself, for fear of being drowned out, finishes up shouting. If someone external comes upon such a situation, he or she gets the impression that it is a crazy situation, although nobody is clinically insane. Note that this is independent of what is said by the speaker or whispered by the audience. In the same way, many conflicts become independent of the initial problem.

The cure to the speaker problem is simple, but not natural: pause. When a speaker stops speaking, after a while, the audience, hearing the difference in the background noise, starts listening to find out what's going on. The same is true in sponsor-CRO relationships: Make a pause, talk not of the inclusion rate, but about the communication. This is useful if done in an objective fashion, avoiding above all the "you started that" temptation, which triggers the "I can prove *you* started it and my lawyer can beat your lawyer" symmetric escalation.

Perhaps we should point out that you do not need training in psychotherapy to manage sponsor-CRO relations, nor do you need therapy to survive it, if some common sense and experience are available.

BACKGROUND FACILITATING CONFLICTS

Sponsors Often Dislike Contracting Out

Outsourcing in clinical development has become a strategic trend within the last ten years in most of the major pharmaceutical corporations. Nonetheless, the CROs should accept that outsourcing is still viewed with an affective bias, not just as another management tool. The rationale was and often still is the belief that the sponsor would do it better in house. It is out of the question to open the debate to determine if a CRO can do a project as well or better than the sponsor (probably "it depends" is the only credible answer), but economic pressure has led to strategic outsourcing of clinical research projects, boosting the growth and consolidation of the CRO industry. And, as in all industries, implementing strategic outsourcing can elicit internal hostility.

CROs may have the nicest and most competent employees and provide the best experience in working together with sponsors, but sponsors' staff will confess at some point that they hate their management forcing them to contract out. Why? Bad previous experience, loss of knowledge on the compound, relationship with clinicians, poor quality, slower development time, high turnover in CRO personnel, and no guarantee of outcome are the typical answers.

What is often not discussed openly is the fear that CROs will take the pharmaceutical jobs away in the clinical research area. This may not be totally rational, but it is almost always in the background.

Line Players' Experience

It should be emphasized that gaps in the project staff experience virtually guarantee a high probability of conflicts when problems occur. Regrettably, there is certainty that problems will occur; this is the nature of clinical research. To prevent problems from turning into conflicts, both sponsor and CRO should involve at least some employees with adequate service experience who will be able to help before a problem develops into a conflict.

Experience includes not only technical and clinical experience, but also business experience (including business etiquette) and service contract management experience. Any lack of previous experience on either side gives rise to too high expectations and thus, to disappointment and to subjective and irrational reactions.

An example is the choice of a CRO based on its sales force's promise that a certain system guarantees faster, better, cheaper patient recruitment. Experienced staff will know that the limiting factor will remain, in most of the cases, the patient-doctor relationship. The problem may be caused here by the expectation elicited by the word "guarantee," even if there are no legal implications.

Internal Politics and Hidden Agendas

CROs often employ "business candid" personnel. Although having adequate technical experience, they often do not know or cannot imagine the internal pressures experienced by the sponsor's organization. One classic example is the international relationship between subsidiaries of a multinational pharmaceutical company. It may well be that there is a greater level of secrecy between the head office and a subsidiary than there is between a sponsor and a CRO. When a candid CRO employee discusses a head office project with the local subsidiary, thinking this will make a good impression, a chain of conflicts may arise. This was not infrequent when CROs were mostly local. Now that many are multinational, they have their own share of internal politics, and thus may understand their sponsors better.

Sometimes the sponsor's agendas are totally hidden from the CRO, which cannot understand what the real problems are. An open and listening attitude is the only way for the CRO staff to understand how to please both the local contact and the central project management team.

Anxiety

As a CRO, we find that anxious sponsors are generally the most difficult ones. This can be either because the expectations are too high (the feeling that this contract with the CRO is the sponsor's last chance puts too much pressure on the CRO). Or perhaps they are simply nervous about everything, and they overreact to whatever problem arises.

For the CRO, understanding the context of the project is vital. Perhaps there is a merger going on. What if both components of the merger have a similar drug in development? What if a promotion is at stake for your client contact? On one occasion, when we heard that two previous studies had shown that the placebo had performed as well as or better than the active compound in a best-selling drug's "profile enhancement" program, we, as a CRO, initially heard the sponsor's "last chance" statement as a great scientific and business opportunity. Then we learned a lot about conflict management.

TALKING ABOUT MONEY

Most people seem to think that the sponsor-CRO relationship would be much better if money was not involved, forgetting there would be no CROs in that case. It is true, however, that money engenders two ways to spoil a sponsor-CRO working relationship:

1. Talk too much about money: All sponsors hate what they view as greedy CROs, who appear to always stick a hand in your pocket, arguing this is not included in their contract. More than bad behavior, this comes from poorly written contracts and poor sponsor (and CRO) staff education. But it spoils the relationship and seldom generates repeat business.
2. Talk too little about money: The relationship is great until a large invoice is presented for all the extra work performed, but not included in the contract. In addition to the problem of how to present management with this large, unbudgeted, supplementary invoice, the sponsor feels cheated and may well refuse to pay, triggering the CRO's retaliation, who could take the study as a hostage. Money is part of the relationship, and some people should talk of it (like contract managers on both sides), guiding their scientific colleagues regarding what is and what is not advisable or acceptable. As a rule, the later the CRO mentions overbudget invoices, the more likely the conflict: The sponsor feels blackmailed.

We will not address the hot issue of the sponsor-CRO contract structure (this is addressed elsewhere in the book). It is enough to say that the CRO would like to be paid for the time spent, irrespective of the outcome, and the sponsor would prefer to pay only for a relevant and completed study. Negotiation allows the parties to set the rules before the contract is signed. If things change, both parties should sit together as soon as possible. And whatever has to be renegotiated should be done as early as possible.

Experienced sponsors may, from time to time during the contract, ask the CRO if they are happy with the sponsor as a client, allowing the CRO to speak up presently, or hold their peace later. And CROs should not allow their perceived need for business to lead them into signing any and all contracts.

Oral Versus Written Communication

As in all business relations, there are two mistakes to avoid when sponsors and CROs communicate:

1. Talking too much, writing too little: There are two main reasons to write. First, it is not always clear what was a chat and what was a decision. "We could do that" is a dangerous sentence. If there are no written approved minutes, pathetic explanations such as "I said this but I meant that" or even disputes like "I never said that" can occur. Especially in cross-cultural settings, what is obvious to one party may carry an opposite meaning to the other. In a meeting between a British sponsor and a continental CRO, the sponsor said, "We are not very impressed with the CRO's performance." The message the CRO received was: "They said it is just okay. Could we improve the performance, even with extra costs?" One can imagine the anger of the sponsor, who was sure he quite clearly communicated to the CRO how unhappy the sponsor was.

   Unspoken assumptions can also cause problems. The statement "I thought it was obvious, we always did this" is a frequent source of negative judgments on either side. Nothing should be considered obvious, at least in recent relationships.

   The second reason to write is that when a dispute arises, only written minutes allow the involvement of "fresh" and experienced staff who might be able to arbitrate fairly.

2. Talking too little, writing too much: Especially in transnational cooperation, some employees are uncomfortable speaking English; some even are not comfortable understanding what they are told. Thus, they write. And e-mail allows writing too easily. Why is that bad? A written statement is perceived as formal, and if not discussed or negotiated, can be understood as an ultimatum. Very few scientists would think of including a disclaimer, like "this is for discussion purposes only." Hastily or informally written e-mails are also dangerous when an insufficient social relationship has been built, because it is easy to run from office to office with a printed piece of paper, saying, "Look how they talk to us. Who do they think they are?" Most basic business training does not teach how to differentiate between humor and sarcasm, and what was intended as innocent casual humor can easily be perceived as an insult.

   A good principle should be "talk first, and confirm in writing later"— never the opposite. It may also be a good idea to telephone to say you are sending some thoughts you put in written form, for discussion only. In some cases, the other party may thank you for this courteous way to proceed by giving some useful verbal information.

   A last and obvious business tip is to consider carefully whom you send copies of written communication to, trying to guess the other parties' reaction before you send them. The danger with e-mails is that it is much too easy to click on names of vice presidents, and the minor irritation may turn into a corporate conflict at the speed of e-business.

DIAGNOSTIC AND PREVENTIVE BEHAVIORS IN AN EMERGING CONFLICT

In this section, we will discuss the sponsor-CRO relationship at an operational level while the study is under way.

Change in Tone

What was a cordial, even familiar, relationship can cool off rapidly. If you are the CRO and you sense a change in tone, you know something is probably going on. If your sponsor has been acquired by another firm, it is important to understand how this can adversely affect your relationship. One of the best leads may come when talking to a client's subordinate or assistant. When the tone changes to aggressive or evasive, you know now what the boss thinks and communicates about you. You may, similarly, receive an offensive or overly critical letter out of the blue. Nothing is better to escalate the conflict than to write an offensive letter in return.

If any change in the relationship is perceived, a request to meet and talk is the best reaction, even if this involves a plane ticket and a day out of the office. Solving a conflict, if possible at all, will cost much more.

Micromanagement

Micromanagement is a key factor causing or presaging conflicts. Micromanagement may simply be a result of the project manager's personality, but most of the time discussions show that the micromanager justifies his or her management technique by a claim that the situation needs to be saved. What is micromanagement? It is sponsor involvement in decisions that should clearly be the responsibility of the CRO. No micromanager will admit "I micromanage, but I can't help it." All micromanagers will insist they are very conscious of quality and there is no such thing as an unimportant detail in such an important project. What the CRO does not hear is the internal discussion between the micromanager and his or her boss. The micromanager complains that, since the project was awarded to this CRO, he or she is much busier than if he or she had done it alone. The reason will be that the CRO is unable to manage the project, forcing the micromanager to attend to all the details to save the project.

A dangerous cycle of dependency can develop. The sponsor will micromanage the CRO. The CRO employees will cede to the client and will step back. They will learn that for every detail, they will call or write the sponsor, asking the sponsor's permission or opinion about everything, as if they were totally incompetent. The micromanaging sponsor will be encouraged by this attitude, which seems to confirm his or her opinion that the CRO is unable to manage the project, prompting him or her to check everything.

There is no way to satisfy a micromanager's anxiety, certainly not by referring all matters of discussion to him or her. Breaking the vicious circle is urgent. If a discussion is not enough, upper management should be involved. A small crisis is better than a big conflict, where the CRO will have lost money (micromanagers can ruin a CRO), the sponsor's confidence, and, sometimes, the whole project.

Lapidation and Diversion

"Lapidation," literally to kill by stoning, is used here in the sense of systematic criticism. The sponsor shows methodically that everything is wrong, bad, demonstrates incompetence, and so on. Generally, the first stones were justified, perhaps because of a big mistake or a misunderstanding of the sponsor's need. But the whole mountain thrown at the CRO, stone by stone, shows both anger and loss of confidence.

Destructive diversion is another tactic used by a sponsor who has made some major mistake. Perhaps the study is very late. However, the sponsor concentrates on a detail that is not essential to the study progress. The worst reaction the CRO can have is to decide that the sponsor will be appeased by an ever increasing amount of detail. This is clearly wrong. This sponsor is very anxious, feels very bad, and expresses hostility, not with the key factors that he wrongly approved, but with a detail. The CRO's two natural but unproductive responses are "it is not my fault" and "this is not important."

Note that for both parties, the hostile person has a purpose: to convince his or her colleagues that the cooperation is disastrous. For this person, one obvious mistake by the adversary or an inappropriate memo is enough, because he or she adds "and everything is like that." Contagion is thus assured.

One solution would be to organize a meeting: Acknowledge the tension, irritation, or anger, whatever is expressed, and refocus on improving the relationship and on the real problem underlying the lapidation or diversion.

Alternatively, the CRO (and also the sponsor) must never forget that hostility is the desire to punish the other party. A CRO can repair a lot by accepting punishment. A hostile sponsor will not become rational again before feeling that the CRO has been punished. The CRO has to weigh the merits of accepting punishment as an investment to defuse the situation, or starting the battle now. Whatever the CRO decides, pride should be left behind: Apologize when there is something to apologize for. This has a soothing effect on the other party's ego. The CRO is better off pleading guilty. The default mistake, even if everything was done right, is that the client perceived poor service, which often was only poor psychology. Note that the previous advice applies with regard to relatively minor issues. Apologizing for ruining the development of a compound cannot be helpful.

Decoding Overly Prudent or Distrustful Attitude

Some sponsors have had a bad experience with a CRO and promised themselves that they would not let it happen again. The next CRO feels that they are paying for the previous CRO's faults, although they have not themselves done anything wrong.

It is important for the sponsor to defuse the situation by being open about the history. If the CRO is aware of that history, it would be helpful to inquire about what the previous CRO did to this sponsor, both as a way to defuse the tension and suspicion and as a way to avoid repeating the problem.

Decoding Silence

One of the most frustrating situations is the "hurry up and wait" situation. But, the CRO may not be aware of the sponsor's internal issues. Sometimes, obvious and urgent decisions are held up without an apparent reason. The temptation to accuse the apparently responsible manager of negligence or incompetence should be resisted.

Instead of repetitive memos, the CRO should call for a frank discussion. Here the CRO can describe the risk to the project caused by the silence and request some explanations. In this way, the minutes of the meeting will show that the CRO did its job, which (eventually) is reassuring for both parties.

Decoding Small Talk

One of the surest internal signs of impending conflict is when, regardless of whether you belong to the sponsor or the CRO, you hear "Oh my God, how stupid/incompetent can they be" statements. If, despite the basic corporate policy of political correctness, there are country-specific negative comments, you know hatred is not far away, and a conflict is bound to arise. Bringing these kinds of comments to light internally can help you defuse the situation. Educating colleagues that "I was only joking" is as dangerous as smoking in a gas station.

Understanding Frustration

A CRO may not sense and adequately deal with the frustration that is often felt by a misunderstood sponsor. The sponsor may, for example, have repeatedly asked for a response to a certain question, to which the CRO, not understanding, has repeatedly declined to provide an adequate answer. Perhaps after a complaining letter to the CRO's upper management, the reply is: "We don't think it is important" or, more evasively, "It isn't in the contract." From the sponsor's point of view, however, if repeated requests are made, it is important—by definition. Even if the response disturbs the busy routine of the CRO, which is carrying out tasks it believes (perhaps correctly) to be more important, a response is nonetheless required.

Developing parallel communication channels can release this frustration. Checking to see how a sponsor's technical employees feel about the cooperation is a safe way to defuse possible conflicts. Some sponsors involve dedicated personnel, like contract managers, and the CRO should also avoid allowing the client to feel trapped in a dead end. Sometimes sponsors are shy or insecure and do not involve their management, afraid that damaging the long-term relationship with the project leader will jeopardize the project.

Several themes recur throughout the preceding advice. These include:

- The feeling of urgency to resolve issues;
- An understanding effort to maintain a win-win attitude and to avoid win-lose situation; and
- Always to envision negotiation as the way to end a conflict or a relationship.

CORPORATE PREVENTION OF CONFLICTS

The following list cannot be seen as a comprehensive checklist of precautions certain to avoid conflicts, since a conflict generally bursts in an unexpected area. But the suggestions that follow can generate a corporate attitude that will be most helpful in ensuring good cooperation with outside resources, thus decreasing the chance of conflicts and their possible impact.

- Sponsors, select CROs you like (i.e., know the people you will work with). If there is a negative feeling at the start of the project, you can see the conflicts flying over the project like vultures.
- Write contracts that minimize or prevent gray areas and allow no room for interpretation. Who does what, by when, is now the rule. It is when parties agree that contracts are signed, not in the middle of a conflict.
- Assess the CRO performance with respect to the contract as well as internal and external benchmarking data. This will help to assess the good and bad and provide a fair and objective appraisal on which management will make decisions.
- Make sure enough time and effort are invested by both parties to cement a human (social) relationship. Take for granted that problems will occur and recognize that it is only during the honeymoon period (when the CRO has won the project but has not started the tough work) that a positive climate can be installed. Then make sure of maintaining that positive climate throughout the project, despite problems occurring, staff turnover, and the erosion of time.
- Make sure that coaching is available on both sides to keep the sponsor and the CRO oriented toward common goals. If the view is that the CRO is only here to make money and the sponsor only here to get registration, good teamwork is not possible. Both the sponsor and the CRO need to create a common success story. Professional recognition is needed for the team on both sides.
- Communication should be an obsession, and the sponsor and the CRO are both responsible. Have personnel been trained? Is there a mechanism whereby dead-end situations are avoided? Are there designated facilitators on both sides? Does everybody on the team, including newcomers, know all of the above? Upper management should be involved from the start, but not too actively. Periodic meetings should allow fast decision making and arbitration, so that conflicts don't occur. In addition, because, for anxious people, no news is not good news, formal and informal exchanges and progress reports should be institutionalized.
- Project staffing is critical. Do the participants form a good team? Is previous experience adequately represented? Are linguistic and technology skills adequately represented? Confidence from the sponsor side will be built when problems are presented with their adequate solutions by the CRO staff. It will be destroyed when unsolved issues will have to be addressed by the sponsor, who will wonder how many other issues are unattended to by the CRO.
- Skilled negotiators on both sides are absolutely necessary. Winning over the other side or punishing them will have negative and costly effects, which are usually avoidable. Conflicts and problems will end well with effective negotiation. All

problems have a solution (including termination) but this solution can be optimal only if both parties are ready to cooperate.

- Because it can only generate conflicts, never reward the following negative feelings:
  - ◆ Greed, which generally has a high cost on both sides;
  - ◆ Pride, unless it is the sponsor's pride in the success of the project, the CRO's pride in the project's success, or the success of the service relation and repeat business; and
  - ◆ Personal agendas. They will always exist, but by assembling the right team, the consequences of personal agendas should not impact the project negatively.
- Training: It is too expensive to reinvent the successful do's and don'ts in sponsor-CRO cooperation for each project. Experienced professionals can share their knowledge, but most of the time, the sponsor and CRO don't have the time, and thus throw fresh troops into the battle, hence the learning curve continues upward.

CONCLUSION

For somebody who has been part of twenty years of contract clinical research evolution, there is a temptation to say "Plus ça change, plus c'est la même chose" (i.e., the more it changes, the more it is the same). It seems to be the same because it is still a very human adventure to complete a clinical study: human preparation with human experts making the best bets for proving the relevant concepts and facts to obtain registration several years later; human cooperation between many investigators, often with cultural and linguistic barriers between sponsor, contractor, and investigators; and the human aspect of the physician contact with patients, which causes most of the unpredictability. But the scene has changed during the last years. What was close to the culture of medical practice (i.e., more of an art) has evolved into a business culture in which time frames are contractual, with bonuses and penalties; quality is required and audited; and business etiquette prevails even over the medical profession.

But whatever way one looks at it, sponsor-CRO conflicts create an infection that, more than ever, can and should be avoided by preventive strategies, systems, and training. When conflict happens, it should be treated vigorously, with courage, honesty, and the project objective firmly in mind. The long-term stakes are huge: the completion of the project in a highly competitive environment.

The industry is ever more striving to reduce the time to market of new drugs and therapies. There is seldom enough emphasis on eliminating delays that are caused by avoidable sponsor-CRO conflicts.

Rakesh Nath

The Role of Site Management Organizations

Site Management Organizations (SMOs) aim to enhance the quality of the patient/investigator interface to improve the clinical trial process efficiency at the investigational site level. An SMO is characterized by the following attributes:

- Owns, leases, or has an ongoing contractual relationship with investigational sites;
- Investigational sites participate in clinical trials with potential therapeutic benefit;
- Manages and assumes responsibility for undertaking clinical research on a commercial basis; and
- Customers are primarily pharmaceutical companies, biotechnology companies, medical device companies, and CROs.

Customer motives for relationships with SMOs, and thus their role in outsourcing clinical drug development, vary, but can be broadly classified as either tactical or strategic. Companies using SMOs tactically 100 percent of the time do not recognize any strategic benefits to building a partnership relationship with these suppliers. Quite often, these companies have already been down the strategic relationship route (via preferred partnerships) with disappointing results. It may be the case that these arrangements were not perceived to allow the required flexibility, since research is not predictable. As one anonymous company official claimed in an interview, "Any decisions below 'this drug needs to be developed' should be tactical, because it needs flexibility and the job just needs to be done."

In contrast, sponsors approaching SMO use as a strategic opportunity largely recognize the potential impact (in net present value terms) on lifetime product value of including SMOs (as a concept) early on in clinical development programs.

GAME THEORY: SMOs AND THEIR INTERACTION WITH OTHER OUTSOURCING PLAYERS

Applied Game Theory (Gibbons 1992) is about value—how to create it and how to capture it. It approaches business not as a war to be fought, but as a game to be played. Unlike war, one can succeed in business without it necessarily being at someone else's expense. Self-interest, however, cannot be ignored and there are times when a win-lose approach is preferable to a win-win approach.

With a Game Theory mindset, business is both cooperation and competition—cooperating with other players to create value and competing with them to capture it. There is, therefore, a dynamic relationship between players in any given market: *co-opetition* (Nalebuff and Brandenburger 1996).

Contemporary Game Theory, when applied to business strategy, focuses on the interplay between competition and cooperation. It does so by providing a framework (PARTS) to break down the complexity of real-world business environments, identifying both the key components and those places where the game can be changed in one's favor:

- Players;
- Added value;
- Rules;
- Tactics; and
- Scope.

Players

The value net

The other players in any organization's market can be described as customers, substitutors, suppliers, or complementors. This type of market map (shown in Figure 8.1) is described as a value net (Nalebuff and Brandenburger 1996) and illustrates the play-

---

Figure 8.1   The value net

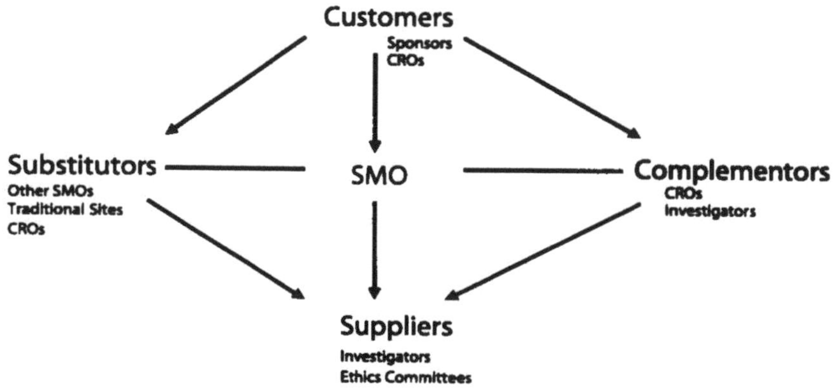

(Adapted from Nalebuff and Brandenburger 1996.)

ers, their interdependencies, and the direction of the money flow (the flow of services and products is in the reverse direction). Every SMO will have its own value net, as each has a different private market.[1]

Substitutors

The term "substitutor" has been used in place of "competitor" as a reminder of where competition really comes from. True competition comes from any offering that customers perceive as a substitute for your own (i.e., capable of the same task or function) and not necessarily organizations that are generally referred to as competitors (largely because of structural or geographical similarities). Some unconventional thinking may be required to fully identify all such sources of competition, particularly those that are currently not competing but may be sometime in the foreseeable future. At any one time, an individual SMO's list of substitutors may include some or all of the following:

- Other SMOs;
- Traditional sites (General Practitioners [GPs], academic centers, nonacademic hospital sites, and so on);
- CROs; and
- Others (e.g., professional recruitment groups and Web-based recruiting).

SMOs (incumbents). The most significant competition for an individual SMO currently does not come from other SMOs, but rather from traditional investigator sites. In fact, some SMOs go as far as to say that, when working on a study in which another SMO was involved, the organization's participation was looked at favorably. An excellent performance by the other SMO was considered advantageous to the respondent and to SMOs in general—a true win-win situation in a nonzero-sum game.

This state of affairs—value creation, in which all players can share—cannot, however, remain indefinitely. Over the last two years, consolidation has resulted in three or four major SMOs and numerous smaller, niche, therapeutic-area players. This will result in a change of emphasis from value-creation to value-capture strategies, with fierce win-lose competition. Competition will occur not only for customers but also for investigators. Figure 8.2 depicts the future shift in attitudes.

SMOs (new entrants). At the moment a player enters a game, it has changed instantly, and is now a different game. The mere presence of a new player has changed it. One cannot interact with a system without changing it. Where there is change, there is opportunity for those who know what to look for. As a general rule, one has to pay to play (fixed assets, professional fees, marketing costs, and so on). Some companies, however, get paid to play by other players who see the value of competition itself. The mere entrance of a new competitor in the market can drive prices down, both as a result of the supply-and-demand effect and when the sponsor uses rival bids to drive an individual organization's prices down. This effect can be intensified if an organization uses aggressive pricing strategies in an attempt to maintain or expand market share.

Sponsor companies, therefore, can benefit from nurturing new SMOs and seeing the number of competing companies increase. Thus, it is not surprising to observe

Figure 8.2   SMO shifts in competition strategies

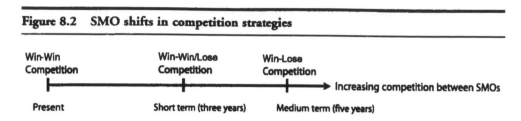

pharmaceutical companies providing financial incentives to SMOs, either directly (start-up capital and so on) or by guaranteeing work to preferred partners. In the long run, these sponsors hope to recoup their investments by lowering the average market price of conducting a study through encouraging fierce competition.

Traditional sites. The only real alternative to using a specific SMO in the minds of most sponsors and CROs is using one or several traditional sites, i.e., independent primary care or practicing, specialist physician practices. These may be sites with which the customers already have established relationships or quite often new, unknown entities that are cold-called.

Although the conduct of clinical trials is most certainly not their prime focus, traditional sites can offer certain benefits over some SMOs, depending on the site's model (e.g., owned sites, GP network, and so on) and how long they have been operating, including:

- Access to real patient data;
- The marketing leverage that comes from working with practicing physicians—the future purchasers of their products;
- Opinion leader status of principal investigators (in academic centers); and
- The benefits of established relationships and experience working with these sites on several protocols.

Although in business terms, nonacademic investigators provide the most serious competition for individual SMOs, in the experience of some SMOs, the most vocally aggressive competitors are, in fact, opinion leaders disgruntled by the loss of ground (and grants) to these commercial organizations. True competition, however, comes from the aforementioned grass-roots investigators (i.e., primary care physicians), and the most significant future threat to an SMO's success may in fact occur at this level. Academic centers are increasingly being perceived by sponsors as slow, costly, inefficient, difficult to work with, and not always willing or able to deliver a level of performance that would justify their premium prices.

In several sponsors' minds, despite the potential to significantly reduce product time to market, the risks of using SMOs exclusively are too high. Traditional sites thus become the true competition for the sponsors' grant money as they are often considered a required part of the site mixture. Senior industry figures, who wish to remain anonymous, have shared the following opinions about using SMOs.

- "As a company we would never just use SMOs, we would mix and match key opinion leaders, hospital based investigators" and other sites.
- We ". . . need to culture opinion leaders, therefore we need a balance between opinion leaders and deliverers of patients (SMOs)."

- "It is not a possibility to just use SMOs because we need to manage risks. It's more risky than just going with a CRO. Only if the timelines or product are not important, which never happens, or we need an unusual group."

This risk aversion culture is particularly prevalent at the clinical research department decision-maker level (e.g., the clinical research manager in charge of the study). This can be true even if senior management has been convinced of the long-term strategic advantages of using SMOs. Some SMOs explained that, although they had managed to convince senior management of the strategic advantage to using SMOs, this message had not, in the words of a senior research manager, "trickled down to the budget-holding decision makers in the clinical research departments."

CROs. In practice, many sponsor companies currently consider CROs as alternatives to SMOs, despite the arguments set forth in the "Complementors" section that follows shortly. CROs can indeed perform the site identification function that SMOs effectively provide. They cannot control and guarantee site performance nearly as well as SMOs, however. Sponsors choosing between a CRO or SMO(s), therefore, must first identify their underlying motive for using an external organization in the first place—simply to identify and enroll sites, or to obtain a measure of control and quality assurance regarding interaction at the patient/investigator interface. In the former case, CROs are indeed a rational substitutor option. The latter case, however, highlights the distinction between SMOs and CROs, weakening the argument for considering CROs as direct substitutors.

In either instance, SMOs will certainly not make CROs redundant, since the latter have evolved into very complex organizations in which patient recruitment is just one activity among many. SMOs, however, provide the key factor in clinical trials—the patients—and represent the final dedicated professional organization in the clinical trial value chain (patient recruitment being the last point on the value chain to have been professionalized). Despite the critical nature of patient recruitment, this professionalization process is a long way behind other departments and functions, such as marketing and purchasing.

It is here we see a fundamental gap in CRO services—they cannot affect the speed of patient recruitment, whereas SMOs can. The added value of the latter, therefore, includes increased numbers of patients at fewer sites, resulting in less variability, improved speed, and lower cost. It is widely accepted that CROs have not been able to consistently reduce drug development time; rather, they have just taken the burden away from pharmaceutical companies. If CROs had been able to compress development timelines, SMOs would not have emerged in the first place, with the promise of not just improving administration but quality.

CROs competing as substitutes for SMOs tend to have built up databases of physicians, just as sponsor companies have. Some have even set up departments dedicated to developing links and systems between investigator sites and their clinical management groups. For the most part, however, neither pharmaceutical companies nor CROs have refined the process to a great enough degree (e.g., ongoing assessments of the sites). They do not, therefore, offer the consistency and control of SMOs, and so deliver less added value in this area.

In the future, however, this may no longer be the case. Several hybrid CRO-SMO models have emerged in the United States (e.g., Collaborative Clinical Research and

Clinicor), with varying degrees of business success. This hybridization process can go in either direction—SMOs crossing over into CRO territory and vice versa. In Europe, Parexel now offers the services of Parexel ClinNet in the UK, a primary care network SMO, which it acquired, and Quintiles has been developing a system of outsourcing study site coordinators to traditional sites. Should many other CROs operating in Europe follow suit and offer true SMO services, competition between SMOs and CROs will, at least in this market, become increasingly fierce. With the entry of these new hybrids, incumbent SMOs will find themselves competing directly at the level of the patient/investigator interface, an attack on their core business area. One respondent explained the driving force behind such a scenario: "When monitoring becomes less time consuming (because of SMO efficiencies), CROs with large overheads will have to move into trial conduct and develop systems of patient acquisition, the most critical part of the development process."

One interesting development is the U.S. Food and Drug Administration's (FDA) review of the potential conflicts of interest inherent in such cases, where hybrids find themselves offering and performing audits, through their monitoring services, on their own study performance at site level. The FDA has, however, acknowledged that pharmaceutical companies do possess their own, in-house Phase I units, which they routinely audit themselves. This issue is further discussed in the "Complementors" section below.

Suppliers

Suppliers include investigators and ethics committees (ECs). In the same way that bringing in new customers reduces their individual added values, increasing the number of investigators or ECs, for example, to which an organization has access increases its bargaining leverage with these players.

Customers

Customers include CROs and direct sponsors, such as pharmaceutical companies, biotechnology companies, and producers of vaccines and medical devices.

As customers, CROs may react very differently to pharmaceutical sponsors because their motives and sources of income differ from SMOs'. Traditionally, much of a CRO's revenue stream has come from labor-intensive tasks, and for the most part, the longer it took, the more the CRO gained (within limits!). CROs could perceive SMOs as sources of competitive advantage and creators of value through increased efficiencies and profitability of functions. For the most part, however, they are currently still perceived as and treated as simple alternatives to traditional sites. Thus, CROs may insist on maintaining their standard level of site monitoring and so preserve their fees for these services rendered on behalf of the ultimate sponsor. This may be the case even though the SMO site efficiencies and Quality Control (QC) procedures have in fact made a large part of the monitoring redundant. Monitoring an SMO is much easier than monitoring a traditional site, because the SMO may well have effectively audited its own work already.

The phenomenon of becoming one's own supplier is also apparent in the SMO sector, and it is widely known that at least one major UK pharmaceutical company is actively sponsoring SMO start-ups. In addition, several other global pharmaceutical leaders operating in Europe are certainly considering the advantages of nurturing such organizations.

The sponsor outsourcing part of its clinical research work to the newly allied start-up maintains the customer-supplier relationship. Should the sponsor maintain an equity stake, this joint venture then demonstrates a backward integration strategy by the sponsor company that suggests strongly its position about where the greatest impact on reducing clinical trial time and cost (long-term) will be in the foreseeable future—at the level of the investigational site and its interface with patients.

Other motives for sponsor companies' participation may include:

- Desire for closer interaction and control of the supplier (SMO), thus reducing uncertainty and risks related to using the organization, including ensuring compatible technology;
- Defense against market dynamism to reduce transaction costs; and
- Leverage greater size to shift the balance of power away from the SMO.

With an understanding of both the advantages and disadvantages for each party in such an arrangement, a win-win situation may be possible—a long-term relationship between a supplier with proven capability to reduce study timelines cost-effectively and a customer whose needs are well-communicated and understood thus met consistently.

Strategic alliances are not always successful and one sponsor's experience of preferred partnership agreements with a network of sites in the United States illustrates this. A senior executive explains: "It didn't work because it was not flexible enough and all our eggs were in one basket—highly risky. Also took very long to negotiate the contracts with SMO central people. Now we would only use them if we have to, i.e., blackmailed by their control of access to patient populations."

Complementors

Complementors include CROs and investigators. Where a company's offering is more highly valued by customers in combination with another company's than it is alone, the organizations are said to be complementors. Thus SMOs offer much more value to customers when they combine their processes with quality investigators. Similarly, renowned academic investigators may offer better value if used in conjunction with an SMO that actually manages the site trial activities and ensures industry-standard quality.

CROs and SMOs can also be considered as complementors under some circumstances. The more successful and numerous SMOs become, by way of their core business activities at the patient/investigator interface, the greater the need for CRO support (data management and so on) of these sites. Similarly, because successful CROs can obtain more studies from sponsors and thus greater investigator grants to distribute with SMO partnering, CROs who use SMOs benefit from this increased business and revenue.

Recognizing the opportunities for increased revenues, some SMOs are acquiring or setting up their own CROs. In doing so, they ensure supply and create competition.

However, as both CROs and SMOs implement vertical integration strategies (backward and forward, respectively), the organizations' relative positions on the value net change. The danger of SMOs attempting to grow revenue by offering a wider range of services to sponsor companies, is that the win-win situation described above is altered by the other side's perception that the expanding organization (or at least part of it) has become a substitutor. The well-publicized financial difficulties experienced by Collaborative Clinical Research are at least in part due to this effect. Its DataTRAK data-monitoring system, for example, was perceived as a threat by CROs.

Covance has publicly stated that although it is very interested in working with SMOs capable of expediting the development process, it will not work with SMOs offering competing (CRO) services. In some SMOs' experience, however, the CROs' territorialism observed two or three years ago is now fading, as CROs sense that SMOs can complement rather than duplicate their offerings. There is no doubt, however, that perceived conflicts of interest may lead to loss of customer (CRO) base.

SMOs offering services such as electronic data handling systems or centralized data processing/cleaning, therefore, may be perceived by CROs as competitors, because the organizations are now operating in the lucrative CRO core business of managing multiple-site data. Other traditional CRO activities that SMOs are beginning to offer include protocol/Case Report Form (CRF) design, ethics committee submissions, medical expert reports, and selection/management of third-party sites outside their own network.

Expansion into monitoring services has a specific, additional complication, over and above the hostile CRO reaction described above. Several sponsor companies see potential conflicts of interest, asking how an organization can audit its own work without bias. SMOs offering such monitoring/Quality Assurance (QA) obviously emphasize that the work is performed by an independent department staffed by individuals with no responsibility for running the study. It must be noted, however, that despite these claims, many sponsors remain sceptical, and so expansion into the monitoring arena may harm core business results. One sponsor who wished to remain anonymous had the following advice for SMOs considering the move into monitoring services: "Leave monitoring till they've grown, until they want to cross the divide [SMO-CRO], because the service is not needed by us from this type of operation— we would instead go to a CRO, with a lot of performance metrics. SMOs should not leave their niche. . . . Sponsors do audit themselves, but SMOs with CRAs [Clinical Research Associate] are moving away from traditional sites. Even if there were two separate businesses or divisions, we would need a seal of approval first, by audit, including their auditing processes, i.e. an audit of their auditing."

Another sponsor had the following to say on this issue: "I don't think it is a good idea [SMOs monitoring themselves]. We would not place a study with an SMO that insisted and didn't let our CRAs in."

Monitoring aside, the core SMO offering, access to investigator sites and their patients, should not be regarded as a direct threat to CROs. The SMO core business, unlike the CRO core business, is not an alternative to a sponsor company's own medical/clinical research department—company CRAs, for example, cannot dose patients or assess adverse events. Instead, SMOs are for the most part substitutes for traditional primary and secondary care provider sites. SMOs offering only core business activities limit themselves to investigator functions. It is only when an SMO begins to offer the noncore activities, such as protocol design and centralized data management, that direct competition with CROs occurs. These supporting activities are indeed functions performed by sponsor company departments, which can thus be outsourced (to CROs).

It has been noted, however, that the very efficiencies in SMOs' core business also reduce the need for monitoring activities. This may result in sponsors requiring (and paying for) less CRO monitoring or even handling all the monitoring in house, thus cutting out the CRO all together. Indirect competition, therefore, is an issue, the significance of which varies between different private markets.[2]

The degree of competition between SMOs and CROs can be viewed as a spectrum as Figure 8.3 illustrates. An individual SMO's position on the spectrum is determined by the amount of overlap between the services offered by it and by CROs. Close to one end of the spectrum, we find a theoretical "pure" SMO, providing only access to the patient/investigator interface. Here, there is no element of outsourcing by the sponsor company, since the activities that occur at this interface cannot be performed by their own staff. Direct competitors at this end of the spectrum comprise only other SMOs and traditional sites. A degree of indirect competition with CROs as described above, however, may occur. This is more likely with an SMO than with a Contract Investigational Site (CIS), particularly because an SMO may also provide a (partial) substitute for CRO site identification activities, by means of the sites already in their network. The actual degree of competition in financial terms, however, is difficult to quantify. Moreover, CRO revenues "lost" to the efficiencies offered by SMOs may be more than recovered in coopetition situations, as illustrated by the PPD Pharmaco-ARC case described below.

The other end of the spectrum represents organizations ("pure" CROs) to which work can be outsourced (i.e., they can perform sponsor company department tasks and functions). Here, organizations face competition from other CROs and (mostly) sponsor company in-house departments. Despite the increasing level of CRO involvement in clinical development, figures reported in *The Pharmaceutical R&D Compendium* (CMR 2000) show that the time to market for drugs has not decreased in the last ten years. Traditional CRO activities alone are not thought to have increased efficiencies or added value at the site level.

A site organization's position on the spectrum has far-reaching implications for how it is perceived in the market by other players (including customers) and for its competitive strategy. Site organizations seeking to sell its services to CROs should obviously position themselves as far as possible from the outsourcing end of the spectrum. They should strive to create the perception in the minds of these customers that the site organization is just a provider of superior site services and not of monitoring/study

Figure 8.3   The outsourcing spectrum

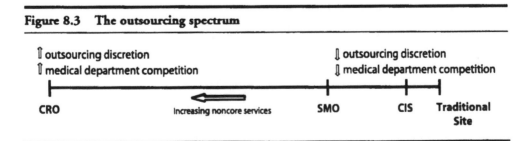

management, thus appearing to be a source of competitive advantage, rather than a competitor to be shunned or attacked.

This can be best visualized in a positioning map, shown in Figure 8.4. An SMO try-ing to sell its services to CROs should strive to create such a picture in the minds of these customers.

Acquisition

While some CROs are sitting on the fence, others are taking action now, having seen the potential added value they can bring to the game by having SMO capabilities. Despite the risk of not being able to channel enough studies to the new acquisitions, several CROs have purchased existing SMO operations outright (e.g., PPD-Pharmaco/CROWN [although the site was later sold again], Harris/MDS, PRA/IMTCI, and Parexel/ClinNet).

One real danger to a CRO-acquired SMO is the loss of business from other CROs. This has occurred even when firewalls are in place to protect data confidentiality.

Organic expansion

Several CROs are creating SMO capabilities from within their existing organizations. Quintiles, for example, is developing not only alliances with Belgian, British, and Dutch investigator networks, but also the ability to provide European sites with study coordinators, as it does in the United States. Additionally, Quintiles supports investi-gators with recruitment efforts and general trial administration.

Supporters of such hybrid models, together with the other alliance/ownership mod-els, view the fast, clean patient data that results from SMOs' access to patients as the key to future success in the marketplace. One possible scenario is the merger of the clinical development market, with CROs acquiring or building SMO competencies and SMOs expanding into full-service CRO type activities. Should this market merger occur, organizations ignoring this trend will find themselves unable to compete effec-

---

Figure 8.4   SMO positioning map

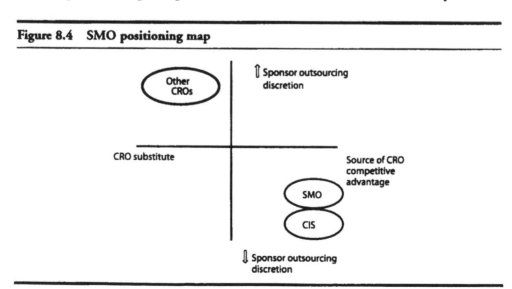

tively in the market of the future. On the other hand, the apparent conflicts involving auditing one's own work may prove to be a more significant market driver. In this latter scenario, competitive advantage is gained not by offering a one-stop service, but rather by ensuring the customer's perception of the organization's independence.

Strategic alliance

Another option for CROs wishing to strengthen links with SMOs without having to acquire them is the formation of strategic alliances. Here, the financial investment and risk involved are much lower and the CRO need not acquire or develop SMO core competences. Some sponsors see strategic alliances between SMOs and CROs as advantageous because specific confidentiality agreements lower the risk of the SMO sharing competitive information with the CRO.

Often a formal alliance comes out of a series of microdeals (i.e., studies conducted) or a preexisting successful relationship between the two parties, such as was observed in the ICON/Clinical Studies merger. Covance, another CRO, pursued strategic alliances and developed partnerships with several United States SMOs. By the middle of 1998, it had preferred partnership agreements with thirty sites, of which five were SMOs.

Some CROs, however, feel that their relationships with and control over traditional investigator sites are sufficient to forego building strong links or alliances with SMOs. They maintain that their core competency in running clinical trials, together with their investigator databases, offer a value superior to that offered by SMOs. Some CROs, like Quintiles, have stated that both SMOs and traditional sites are needed to make up the mix of sites required for success in larger trials. Because CROs do not retain in house certain documents related to their investigators, an SMO offering stewardship of records required for source data verification gains another competitive advantage over CROs.

Win-win-win

As SMOs become increasingly successful, CROs will find themselves doing less site selection and monitoring. Viewing this as an opportunity for value creation and growth, some SMOs are working toward building equal partnerships with CROs, rather than accepting the traditional customer-supplier relationship. In this situation, the alliance makes joint bids to sponsors, with the SMO providing the front-end services (at the patient/investigator interface) and the CRO supplying the back-end support. Such a package is said to be of superior value to a CRO working alone with fragmented sites. Most sponsors, for their part, see no problems with these strategic alliances, viewing the two parties as different, but complementary.

In partnerships such as the much discussed PPD Pharmaco-ARC (Zonagen) alliance (Lightfoot et al. 1998), the revenue ceded by CROs as a result of using an SMO is made up by the increased efficiencies of using an SMO and by the additional marketing clout of a demonstrated ability to provide access to investigators and patients. Similarly, inSite, a United States SMO, reported that it is in discussion with several CROs about possible alliances, despite the fact that the SMO already offered project management and data cleaning services.

The PPD Pharmaco-ARC collaboration demonstrated how all involved parties (including the sponsor) can gain in a win-win-win situation (i.e., in order to win, somebody else does not necessarily have to lose). As the CRO-SMO team was in place at the beginning of the project, it was able to accelerate every aspect of the study process:

- Clinical development start-up;
- Enrollment into the two pivotal trials;
- Data cleaning; and
- Database locking.

The successful delivery of the accelerated timeline benefited each player.

The *sponsor* benefitted by leapfrogging over its main competitor (Pfizer) in the race to be first to market (i.e., a small company was able to outperform a much bigger competitor by the use of innovative methods and by changing the rules of the game). Furthermore, it claimed that it did so more cost-effectively than Pfizer. Although on a per year basis, the costs of the trial program are higher than its competitor's, Zonagen saves money in the long run by having to bear development costs over a much shorter time frame. In this case, the sponsor leveraged the CRO-SMO alliance as a competitive advantage.

The *CRO* benefited from the collaboration through increased project profitability, despite reducing project management fees from, say, three years to one year, despite the performance of some front-end tasks (e.g., protocol/CRF design) by ARC and not PPD Pharmaco, and despite no site selection fees. These "losses" were more than off-set by the efficiencies that ARC delivered (e.g., the CRO retained all regulatory activities and ARC ensured delivery of all critical documents not only in a more timely fashion, but also in a cleaner state). PPD Pharmaco would not have been able to replicate this level of performance at a lower or equal cost. Implementing process innovation by using an SMO, the CRO increased its own profitability. Other benefits the CRO gained as a result of the alliance include:

- Potentially increased the size (number of sites) of the contract it would have otherwise obtained, perhaps locking out competitors;
- Marketing leverage gained through publicity of the spectacular results it helped achieve;
- Strengthened relationship with a satisfied customer; and
- Leveraged core competencies of the SMO.

The *SMO* benefited from the alliance by obtaining new business, since it would not have been able to secure such a sizable contract on its own. Like the CRO, the SMO was able to lock out competitors, gain marketing leverage, and strengthen its relationship both with the CRO and with the sponsor. The SMO leveraged resources of the CRO while at the same time creating the perception by other CROs that it is a source of competitive advantage, not a competitor.

Thus, from a market-oriented perspective, by joining forces, CROs and SMOs may offer more value to sponsors than by working individually. This point has been highlighted by several sources, including the ARCP's White Paper on Future Trends (Lightfoot et al. 1998).

Added Value

In the eyes of sponsors and CROs, some or all of the following puported advantages of SMOs may offer added value:

- Fast recruitment through proven patient resources (e.g., databases, lists), processes (e.g., advertising, centralized recruitment centers), and ability to efficiently transfer effective techniques between sites, or even to replace poorly performing sites when necessary (from within the site network, possibly with no extra cost to the sponsor);
- Better subject retention, since the investigators know from experience how best to select appropriate patients;
- Monitoring and training efficiencies—sites are staffed by professionals experienced in Good Clinical Practice (GCP), resulting in time and cost savings through:
  - fewer protocol violations;
  - rapid receipt of clean, high-quality data;
  - faster closing of database;
- Project management efficiencies, since some SMOs provide a central coordinator as a single point of contact for the sponsor and/or CRO. This individual coordinates communication in both directions with all investigators/study site coordinators involved in the study. The added value largely comprises consistency of site performance and information;
- Standardized safety communication to and from sites, via a central SMO coordinator; and
- Ensuring participation of investigators/study site coordinators in the investigator meeting.

It must be noted, however, that the SMO industry does not by any means attract universal praise for bringing added value to the clinical trials arena. Many sponsors, having worked with such organizations, dispute whether they do. The most common complaint is that SMOs increase communication problems with sites, without controlling study conduct.

Repeated poor performance by even a few SMOs, therefore, may have a negative impact on the entire sector, leading to value erosion and decreasing profit margins for all players. One sponsor representative in the UK stated that "our experience with local SMOs is that they have not demonstrated any benefits (except a couple of times, e.g., accelerating patient recruitment rates)." Another had the following to say on the issue: "It doesn't take long to know if a SMO is delivering on promises. If you get a pattern across several SMOs, it affects the view on the SMO concept itself. Therefore SMOs may have to defend their reputations against other SMOs and differentiate themselves from them."

The first respondent quoted above offered an explanation for the lack of demonstrated added value that suggested the SMOs in question were not wholly at fault. He stated that perhaps the sponsors themselves had not allowed the SMO to operate optimally, since they tended to outsource in a reactive rather than strategic manner. He did, however, go on to say that, in his experience, SMOs, as they are "very new," tend to be too flexible and to make promises without having the expertise or facilities.

The challenge for SMOs faced with negative customer perceptions is obvious—to prove their worth by delivering on their promises and building up demonstrable added value in the form of historical and real-time performance metrics.

Rules

The rules of business come from culture, contracts, and law. As in other industries, the rules governing the SMO sector can be a source of power in the game, as they structure the way it is played. However, the rules of the game can be changed—as poker players used to say in the old Wild West, "a Smith and Wesson beats a straight flush." Moreover, new rules can change the very game itself.

Rules cannot be stepped outside or ignored, if to do so risks legal or financial penalties. The majority of rules that can be changed are those concerning negotiating practices and contracts, both with customers and suppliers.

Scope

Boundaries defining games are artificial. For simplicity's sake, we create these boundaries to try to analyze and understand games. To take any one game in isolation, however, is dangerous, as we may mistake a part for the whole. All games are linked to other games, and an event or action in one can affect others.

Thus a successful, ambitious European SMO, *XYZ*, plans expansion into a new geographical market, the United States. It has given careful consideration to the level of competition it will face there, but feels confident in its capabilities and the added value it can offer to customers. Furthermore, its current European operation enjoys a dominant market position and is generating an enviable profit, which provides more than sufficient financial resources to fund the transatlantic expansion plans. Shortly after beginning its foray into the United States market, however, the fictional SMO is in dire financial trouble, with huge cash flow problems stopping its U.S. growth in its tracks. The previously free-flowing capital from its European operation has been reduced to a trickle. Without this source of cash, its expansion into the U.S. market looks doomed. What could have happened?

Ignoring links between games can lead to just such a scenario. The European SMO made the mistake of looking at the United States and Europe as two separate games. It did not see the links between them and how its actions in one game would lead to repercussions in the other. Moreover, it compartmentalized its competitors into those in the United States and those in Europe, with no overlap between games. In other words, it perceived that it was fighting two separate battles on two fronts far from each other.

If *XYZ* really does possess excellent capabilities and could pose a competitive threat, incumbent American SMOs are unlikely to stand by and do nothing. Returning the complement and entering the potentially lucrative, relatively immature European home market of *XYZ* may be the best way to counter its entry into the United States. In the United States, their core market, any price-cutting strategies would be most damaging to the incumbents, hitting their bottom lines hardest, while *XYZ* weathered the price war using profits from its home market to support its relatively small United States cost structure. By entering Europe, however, the United States SMOs have suddenly changed the game(s). *XYZ* no longer enjoys its virtually uncontested dominant position in Europe. It now faces competition from experienced United States SMOs with very deep pockets. Should the latter decide to begin a price war in Europe, *XYZ* would stand to suffer most, and the profit stream that its United States expansion plans depended on could soon dry up—the game(s) had changed for the worse as a consequence of retaliation in one market by incumbents in the other.

The scenario described above can also occur in the event of expansion into new product lines. Thus, SMOs considering expanding into traditional CRO activities such as monitoring and data management might be well-advised to expect CROs to retaliate by acquiring or developing their own site management capabilities.

Additionally, and potentially even more harmful, is the perception among some sponsors that SMOs cannot be expected to monitor their own study performance in an unbiased fashion. Thus, expansion into the monitoring services game may lead to a loss of customers in the core business site activity game.

RELATIONSHIPS

Sponsor-SMO Relationships

The critical nature of a good relationship between a sponsor and an SMO is obvious. The very reason for the emergence of SMOs (patient recruitment) is the first and, by definition, the most significant area affected by the strength of this relationship.

Open discussion of recruitment ideas and expectations has the following consequences:

- Increases creativity;
- Builds motivation and enthusiasm for the project;
- Better communication throughout the study's duration;
- Faster resolution of problems when they arise; and
- All parties benefit and learn from each other's knowledge and experience.

Active relationship building is crucial to creating a solid foundation on which to work. This activity should form an integral part of any clinical trial program. Examples of relationship-building activities that should be budgeted for include:

- Sponsor-SMO investigator meetings, which should have time set aside specifically for a discussion of recruitment planning; and
- Process(es) to promote consistent approaches to and performance of recruitment across sites, by working with investigators.

Historically, inefficiencies at investigational sites occur largely because industry-standard/GCP study conduct is not the site staff's primary concern. The onus has, for the most part, rested with the sponsor monitor to ensure adequate compliance with study requirements. Processes designed to help the monitor's task usually include prospective measures for control over patient recruitment and data handling, and retrospective accounting for discrepancies and errors. SMOs offer significant efficiency improvements at the site interface, thus reducing the need for and costs of sponsor monitoring and quality control. The value of building good relationships, therefore, can be quantified financially.

The role of a monitor has grown to include the business aspects of managing relationships with SMOs. As service providers, these organizations are more customer oriented than the traditional investigator sites, and monitors with effective team management and motivational skills are now in a position to leverage these qualities to optimize efficiencies at the site level. In addition to improving study performance, the monitor stands to personally benefit from the effects of improved relationships with SMO staff by doing away with much of the frustration caused by dealing with inefficient, unpredictable sites and their sometimes opaque methods.

Furthermore, as the increased efficiencies produced by good relationships and effective communication with customer-oriented SMOs reduce the protocol-specific, operational pressures on monitors, they can commit more effort to product commercialization. Strong relationships with investigators facilitate the distribution of product information, providing a platform on which to base future commercialization tactics.

Sponsor companies face both opportunities and challenges at the sponsor-SMO relationship level. Monitors, given their process orientation, are likely to require development and training in order to take full advantage of both the operational and marketing advantages that have become apparent with the increasing use of SMOs. However, the added value and competitive advantage generated by such efforts will be recognized and captured by forward-thinking, innovative sponsors.

The experiences of one SMO operating internationally provide an interesting contrast to the position described above. Although successfully working with a number of sponsors in several other European countries, in one country, this organization found itself perceived as a threat by the local subsidiary staff of these same pharmaceutical companies. Some sponsor staff (and SMOs) see SMOs' main value as their ability to optimize the use of CRAs, ultimately allowing the reduction of their numbers (perhaps by 80–90 percent). This was the prevalent belief amongst CRAs in this particular country, which resulted in their unwillingness to work with the SMO. The SMO was, in this instance, not viewed as a source of competitive advantage or a partner to work with, but rather as a threat.

SMOs may find that performing poorly on one study very swiftly undoes the goodwill and reputation built up over several previous studies. Furthermore, sponsor companies may well be too big to have meaningful relationships with SMOs. The size and complexity of some pharmaceutical companies result in multiple contracts with different people in different departments, and an SMO may find itself having to explain who it is again and again. On the other hand, the positive effect on an SMO's bottom line of investing in relationship-building efforts was described quite clearly by one sponsor who said, "If they have a track record with us, they can almost name their price. It is unusual that we don't go with a SMO after a good relationship has been built up, just because of price."

In an attempt to control costs, many sponsor companies have set up central contract departments that may be able to veto, but often not choose, the SMOs to be used. As far as many SMOs are concerned, however, this attempt to professionalize purchasing is, unfortunately, a policing and not proactive activity in practice. Obviously, the opportunities to build a value-added relationship under such circumstances are limited.

In the long term, mutual respect and working together as partners can create the most value between a sponsor and SMO. For the sponsor-SMO relationship to evolve into a true partnership, both parties must develop a deep understanding of what they, themselves, are trying to achieve and what they need to reach these goals. These needs include:

Sponsor

- Precise and transparent communication of the sponsor's needs;
- Responsiveness to queries; and
- Complete understanding of cost structures.

SMO

- Bid at levels that allow for all required resources;
- Risk assessment excellence; and
- A complete understanding of the sponsor organization's motivation and politics.

As the relationship grows stronger and each partner becomes more confident in the other's capabilities, greater trust is built. At this stage, to best take advantage of the efficiencies that SMOs offer, sponsor companies may need to change their own processes. Granting their CRAs the flexibility required to take advantage of the SMO's efficiencies may in fact necessitate amending SOPs. Most SMOs are frustrated when sponsors' monitors are required to follow company policy despite there no longer being a need for this level of involvement. Potential cost and time savings identified by senior management are therefore not taken advantage of, to the detriment of both sponsor and SMO. Senior management rarely think of SOPs as hampering flexibility at an operational level. As one SMO representative explained, "SMOs should not be treated like normal sites, i.e., there is no need to contact individual investigators, just the central SMO. . . . [N]ot enough sponsors change SOPs when dealing with SMOs. . . . [S]enior management do not think of SOPs but CRAs cannot make decisions flexibly because they do not have the power and are forced to monitor each site individually. Therefore, the key is changing the SOPs."

Sponsors should note that their communication and monitoring skills are more highly valued by SMOs than their financial and scientific know-how (CenterWatch 1998). Since SMOs want to work with professional staff knowledgeable about both the local medical and general culture, a local sponsor office can be extremely valuable. From these offices, medical and monitoring staff with local knowledge can readily foster good communication with investigators/study site coordinators. Sponsor staff are on hand to respond to any questions that may arise during the course of study and can nurture the relationships essential to the optimal conduct of any clinical trial.

In addition to sufficient information about patient needs, to maximize the opportunities made possible by good relationship management, sponsors need to allow the SMOs adequate lead time. Without sufficient notice, SMOs are only able to accept studies as suppliers, not as true partners. This lack of lead time tends to be passed down the value chain, and is thus exacerbated if a CRO is placed between the sponsor and SMO.

Another issue that sponsors may need to address is the manner in which SMOs are remunerated. For the most part, CROs are permitted to bill on an hourly or daily basis. SMOs, on the other hand, are expected to bill on a per-patient-recruited basis, effectively a fixed price contract. Depending on their recruitment process efficiency in practice, organizations either profit or lose if time or costs turn out to be lower or higher. The uncertainty involved can be harmful to any relationship-building efforts, particularly when unanticipated and unattributable problems arise.

The experience described by one U.S. sponsor acts as a useful case history on how not to build a successful relationship. The sponsor's representative explained, "It worked out badly, and we walked away. The SMOs were complacent, bureaucratic and did not deliver. If they got their act together, a sponsor would not need to negotiate twenty contracts with twenty investigators, and would therefore save time. But in reality, the SMOs' hardball negotiating meant that the negotiations took longer than going to the individual investigators. The SMOs' approach was bad, and it didn't feel good even if they delivered, because we felt like we were blackmailed by the service provider." Another sponsor's comments demonstrate that the same mistakes can be quite easily made on both sides of the pond. Of European SMOs, this sponsor says, "We have a negative view of them, but as they deliver we use them. We have trouble dealing with them."

CRO-SMO Relationships

CROs' opinions of the value offered by SMOs can be described chronologically.

As Figure 8.5 shows, the general relationship between CROs and SMOs has evolved. SMOs were largely ignored by CROs at first; then they were increasingly viewed as a threat; and, most recently, they have been viewed by some of the more innovative, forward-thinking CROs as potential sources of competitive advantage. As one pharmaceutical executive put it: "CROs are waking up to the fact that investigators/SMOs control 50 percent of the trial, and this leads to power and value."

The PPD Pharmaco-ARC example discussed previously indicates the type of performance and resulting financial benefits that are possible under optimal conditions. Here, both parties considered the other an equal partner. Furthermore, team-building and communication skills were critical factors—the team had to stay together during the entire development process.

Traditionally, sites have always preferred working directly with sponsor companies rather than "via" CROs. Sponsors that build direct relationships with investigators and retain control of project management are the highest-rated clients. Both sponsors and SMOs rate the presence of professional staff on both sides. From the perspective of sponsors, the advantages of having such professional staff on sites are obvious. For the SMOs, the professionalism sought in sponsor monitors includes good communication skills and adherence to scheduled meeting times. CROs' monitors' training programs should, therefore, ensure that they understand the importance of creating and maintaining good relations with SMO central and site staff.

Sponsor/CRO–SMO Relationships

In the long run, the SMO industry will mature; technological advances and globalization will drive the sector toward commodification at the tactical end. At this stage,

Figure 8.5   CROs' evolving view of SMOs

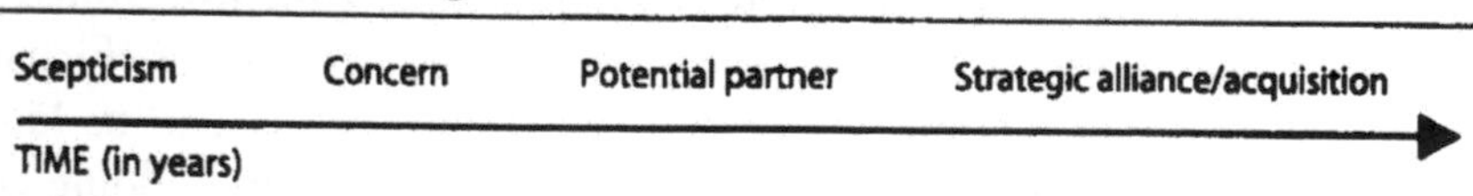

Scepticism	Concern	Potential partner	Strategic alliance/acquisition

TIME (in years) ➤

both CROs and SMOs will find, as Figure 8.6 demonstrates, that it is the "softer," people-oriented business strategy components, including training and knowledge management, where significant competitive advantage and differentiation are to be found within a strategic relationship.

The results of one of CenterWatch's (1998) surveys of investigators are summarized in Table 8.1. Although this survey was based on all types of investigator sites and thus was not limited to SMOs, the issues raised are indicative of customer perceptions of SMOs. These results should be taken into account in any attempt to strengthen a relationship between a CRO and an SMO.

The SMO study site coordinator/investigator–CRO/sponsor monitor relationship is critical to a study's success. It is prudent, therefore, to take certain prospective steps at this level of interaction to create a base on which a stable, productive relationship can then be built. To begin with, the CRO/sponsor must clearly communicate its expectations, required deliverables, and timeline to the SMO. The responsibilities of both parties should be established at the outset. An agreement on a system of measuring and tracking data queries and responses is also highly recommended. These targets and responsibilities are then formalized in a contract, which may include clauses related to:

- Recruitment timeline;
- Definition of an evaluable patient;
- Target number of patients per site;
- Reporting and regulatory responsibilities;
- Report frequency;
- Termination of contract terms; and
- Payment terms.

Additionally, the monitor must be well-trained, understanding the value of SMOs as partners in the trial and responding quickly to their queries. Finally, the SMO study site coordinator/investigator must be suitably trained, experienced, and committed—the monitor must see a high level of attention to the study, indicated by a minimal number of mistakes in CRFs, and so on.

Figure 8.6   Business components that determine differentiation

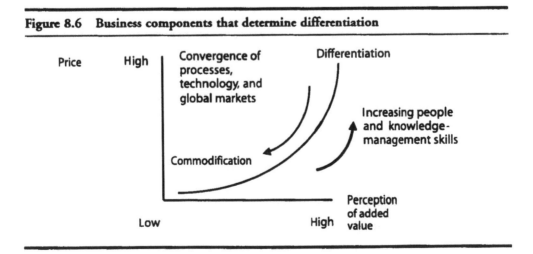

Table 8.1　Survey of Investigators

Characteristics of a good relationship		Characteristics of a poor relationship	
Professional/knowledgeable staff	41%	Incompetent/unprofessional staff	33%
Quality monitoring	34%	Poor monitoring	31%
Good communication	32%	Poor communication	20%
Responsive	19%	Arrogance, lack of respect	17%
Accessible	19%	Unresponsive	16%
Well-designed protocols/CRFs	17%	Poorly designed protocols/CRFs	15%
Prompt and fair payment	14%	Inaccessible	15%
Mutual respect	14%	Payment delays	14%
Realistic timelines	10%	Inflexible	11%
Flexible	7%	Unrealistic timelines	10%

(CenterWatch 1998.)

During the course of the study, minimum standards should include:

- Reevaluation at regular intervals of the original expectations and responsibilities. Changes should be made where necessary;
- The CRO/sponsor should make its best efforts to keep the same monitor involved for the duration of the study; and
- Effective two-way communication channels must be maintained, allowing the flow of both positive and negative feedback.

NOTES

1. The term "private market" simply acknowledges that any given SMO's realistic potential customers are not *all* sponsor companies, nor are its actual competitors *all* the other SMOs, traditional sites, and CROs. Furthermore, the list of organizations actually purchasing services from or competing against a specific SMO is unlikely to be identical to that of another SMO. In other words, we differentiate between the SMO market industry as a whole and the *private market* of individual organizations.

2. One sponsor representative stated that she would not decrease the number of monitoring visits made to SMO sites, nor would the time required for source data verification be reduced. She did, however, concede that the visits should be "easier" (and so less time consuming), giving the CRA an opportunity to add value to the trial process at the site level by "helping the investigator with patient search techniques; premarketing activities; helping to promote the company's brand."

REFERENCES

CenterWatch. 1998. European sites rate sponsors. *CenterWatch Newsletter* 8(6).

CMR International. 2000. *The pharmaceutical R&D compendium*. London: CMR International.

Gibbons, R. 1992. *Game theory for applied economists.* Princeton: Princeton University Press.

Lightfoot, G. et al. 1998. Faster time to market—ARCP's white paper on future trends. *Applied Clinical Trials* 7(4): 56.

Nalebuff, B. J. and A. M. Brandenburger. 1996. *Co-opetition.* London: HarperCollins Publishers.

Patricia Lobo

Clinical Trial Supplies— The Vital Link

Major pharmaceutical players strive to bring their next breakthrough medicinal product to the marketplace, ever conscious of the development costs involved. However, the hurdles to proving safety and efficacy present a formidable challenge, not least because of the numbers of patients involved in clinical trials. Medical statisticians may insist on matching placebos, proper randomization, and double blinding, while regulatory agencies expect high-quality standards for trial materials. Supplies of clinical trial medication can, therefore, become a critical factor in determining the successful outcome of a trial. The special skills needed to produce clinical trial supplies (CTS) add an extra dimension to the complex business of clinical research. The role of a contractor in providing clinical trial supplies services is increasingly important in the area of clinical development. Here, we look at the complex business of clinical trial supplies and how contractors are responding to the challenge of meeting a growing demand for outsourcing the essential test articles that are the very focus of clinical research trials.

WHAT DO WE MEAN BY CLINICAL TRIAL SUPPLIES?

CTS are not restricted merely to investigational medicinal products, for example, novel therapeutic substances that need to be investigated in clinical trials for efficacy and safety or for bioequivalence. CTS encompass an entire portfolio of medication and related items needed for patients or healthy volunteers taking part in clinical trials. This portfolio could include the new active substance, the placebo, the comparator, or a combination of drugs in any standard or special formulation, either alone or with special drug delivery devices. The test articles must be put together into a patient kit, along with all the instructions and patients' leaflets explaining how to use the medicines in the correct regime for the correct trial purposes. The patient kit also includes all other medication, such as supply rescue medication (for use if the patient trial medication is not working), diagnostic kits, and any other medication that a patient may need while on that trial.

Every patient kit is unique. Manufacturing clinical trial materials includes components that are quite different from the manufacture of different dosage forms for commercial purposes, which have different sizes, color, or packaging for different products or different strengths of the same product. Clinical trial materials involve the manufacture of a new medicine, a placebo, a comparator, and/or different doses of the same active ingredient in a similar way (i.e., dosage forms with the same size, color, and packaging so that they are indistinguishable). This identical appearance is essential in a blinded trial; therefore, not only is the manufacture a challenge, but labeling of the product is important as well. Also, the manufacturing process has to be controlled rigorously to ensure that there is no crossover or mix-up of test article, placebo, or comparator. Another challenge is the procurement of an active comparator. Sometimes this is not readily available; therefore, sourcing a supply of the compound can lead to delays in preparing the patient kits.

OUTSOURCING

It is possible to outsource all or part of an entire clinical trial supplies process. The scope covers, for example, management or improvement of the process, including cycle time reduction; the systems used; the development of the systems; pharmaceutical development; the manufacturing, packing, labeling, printing, analytical testing, batch testing, and release of medication and related items; and the distribution, logistics, information technology (IT), tracking, storage, shipment, customs requirements, reconciliation, and destruction of supplies.

Outsourcing can occur on a tactical or strategic basis. For most clinical studies, pharmaceutical companies frequently outsource CTS tactically. A client may require labeling at short notice, but may not have the internal resources to get this done in time for the start of the study. Or, a client may not have the staff to pack supplies for a particular study, so a contractor is sourced to do the job. Strategic outsourcing would involve a decision that distribution or packing or any other clinical trial supplies activity is not a core competency and it should be outsourced to a contractor with that expertise.

Principals involved or potentially involved in setting up single or multinational studies need to assess the local or global capabilities of the various Clinical Trials Supplies Organizations (CTSOs) that offer a CTS service. Some of the items to consider would be: the number of sites these organizations offer; the staffing, capabilities, and core technologies available; the depth of experience and expertise of the staff involved in the particular task required; and the qualifications of key management executives.

The client needs to feel comfortable and assured that everyone on the team, on both sides of the relationship, will have similar technical expertise and depth and breadth of experience. One would need to assess the distribution systems and to be assured that these organizations talk to one another on a regular basis. Just because some CTSOs operate globally does not automatically mean that the subsidiaries from each country operate in the same manner or that there is good communication between them. The subsidiaries need to work in harmony to provide any real benefits to potential clients. In working with a global CTSO, clients can optimize their overall clinical process and can save a substantial amount of time and money since these CTSOs can coordinate CTS activities very efficiently and effectively.

CLINICAL TRIAL SUPPLIES ACTIVITIES

Before considering outsourcing, it may be useful to look at the various activities involved with clinical trial supplies.

Management of CTS

The clinical trial supplies process requires proper management of both the design process and the operational process. In the design process, it is necessary to forecast the demand before designing and organizing the preparation of the CTS.

In forecasting the demand, a number of questions that impact on the clinical supplies area need to be answered, including:

- What dosage forms, different strengths, and frequency of dosing are required?
- What is the shelf life of the supplies and does the shelf life need to be expanded during the study?
- Is the study going to be open, closed, blinded, or double blinded?
- When is the study likely to start?
- Where will the centers be located?
- What will the recruitment rate be at each center?
- How will the recruitment rate change with time?
- What is likely to be the peak number of supplies?
- How soon will it be possible to obtain regulatory approval?
- What are the customs export and import regulations in the country where the supplies need to be distributed? and
- What logistics are needed to get supplies to the patients or investigative centers?

Forecasting the demand should continue during the trial, since a number of items in the design may have to change. For example, the recruitment rate at some investigator sites may not be as expected for various reasons. Consequently, other centers may need to be involved in the trial.

The following activities are crucial elements of the overall operational process:

- Organize the manufacturing and packing of supplies within the supply chain;
- Complete the patient kits;

- Organize the distribution of supplies to the patient or healthy volunteer;
- Provide ongoing supplies;
- Reconcile the supplies;
- Keep a record of the appropriate retention samples;
- Check shelf lives and stability of CTS; and
- Account for the destruction of supplies.

The aim is to get the earliest possible study start with minimum waste in the supply chain, consistent with the earliest possible study finish.

Manufacturing

Clinical trial supplies manufacturing includes standard-type dosage forms, for example, tablets, capsules, powders, granules, and liquids, or special types of dosage forms, such as steriles, lyophilized powders, soft gel capsules, and soft chews. In most cases, it makes sense to outsource the manufacture, not just because the contractor is able to reduce costs (because it can help switch fixed costs into variable costs), but also to save time and because a contractor offers specialized expertise.

Sometimes pharmaceutical development and scale-up of certain clinical trial materials may be required (for example, of biological material), and the contractor may have the facilities, equipment, and the expertise to produce it. Unless a client includes in its portfolio the capacity to handle controlled drugs (e.g., opiate narcotics), it may be better to consider outsourcing. Applying for licenses, filling in forms, and dealing with quota systems and security systems that need to be in place to handle controlled drugs are both time-consuming and expensive. The same applies to the manufacture of penicillins, cephalosporins, and contaminated or other difficult-to-handle products. These often necessitate considerable equipment cleaning and thus slow down or even halt production. A contractor should have the flexibility and ability to respond very rapidly and put up an additional shift very quickly. Some of the manufacturing support activities, such as analytical testing, stability testing, batch testing and release, regulatory affairs, quality assurance, warehousing, storage, and distribution, may be provided by contract manufacturers.

Packing

The CTS have to be packaged in the most appropriate way to facilitate compliance and ease of use by the patients. Therefore, the packaging design has to take this into consideration and balance it with the required ease of the manufacturing and packing processes. Patients are always curious to know whether they are on active medication or placebo. One pharmaceutical company is said to have recalled some of the CTS, because there happened to be a red dot on some of the packaging of the trial material. The patients believed they were on the active drug, when in fact the red dot was due to a marker pen when the printing was done. Various types of packing can be used for CTS purposes, for example, bottles, blisters, cartons, inhalers, tubes, pumps, vials, and syringes. Some of these can be standardized and automated, but many particular patient kits have to be manually assembled and/or hand packed.

Labeling

In the clinical supplies arena, instructions on clinical materials must be very clear to a patient. In the last ten years, there have been significant changes in the automation of the clinical trial supplies labeling process. Clinical trials often require multilabel panels on one container, especially for clinical trials that require multilingual panels. For example, in the pattern-coated label, the first panel is permanently fixed, containing the directions for the patient and the coded patient's number. The second panel contains the same information but it can be removed and attached to the case report form (CRF). The third panel contains the medication's identity to be sent to the sponsor/investigator. The information in the third panel is hidden within a pocket, built inside the overprinted label stock. The back pocket has an on-press application chemical coating so that when pressure is applied to the top of the pocket, the coated-back sheet and chemical coating interact to produce a readable image. Due to advancements in technology, other types of labels have been developed, including the expanded content label (ECL), scratch card label, tamperproof label, and multilingual label.

Previously, labels were printed in the language of the country where the trial was carried out and sometimes they were printed in English because that was the language accepted by many countries. However, clinical trials are becoming increasingly global in nature as more and more countries are becoming involved, both for regulatory reasons and to access the maximum number of patients as quickly as possible. Sometimes, it may be easy to anticipate the countries where the trials will be conducted, but quite often, it is not possible, especially if country or site recruitment rates are fluid or unknown. There may very well be one or two further countries that one may wish to include on short notice during the study. Some contractors have set up multilingual labels to increase the flexibility of supplies. A multilingual label is one that can be used on clinical trial materials with information in more than one language, allowing that particular supply to be used by patients in more than one country. Multilingual labels reduce the number of different labels required, thus reducing packaging and therefore the packaging runs, and the follow-up or the lead time for follow-up supplies. Multilingual labels also reduce waste, achieving an overall reduction in study medication and costs. It is worth noting that the supply of active substance may be strictly limited in Phase II studies; thus its conservation may be a critical issue.

Associated with advances in labeling are advances in the different types of adhesives (permanent and removable). Adhesion is one of the most important physical properties of any clinical trial label. Unlike most product labeling, clinical trial supply labels must occasionally perform under extreme environmental conditions. Usage and storage conditions of labile drugs are critical when making a labeling decision; for example, if the drug must be refrigerated, then special adhesives may have to be used. Labels must also conform to the container designated to contain the study medication (for example, glass or plastic bottles, high- or low-density polyethylene bottles) and should meet the labeling and storage protocol and fall within the label supplier's capabilities (21 Code of Federal Regulations 312 1999). There is also a choice of printing materials (for example, Electronic Data Processing [EDP], plain or premium laser, plain or opaque litho, and plain or opaque thermal).

Storage, Warehousing, and Distribution Logistics

To ensure safe and on-time delivery of CTS, important elements relating to shipping should be addressed during the planning stages (depending on whether the CTS require local distribution or global distribution) (Carmody 2000).

The main issues to consider are:

- Documentation—Release certificate, ethics committee approval, regulatory approval, Good Manufacturing Practices (GMP) certificate, import license, pro forma invoice, customs invoice, and acknowledgment receipt.
- Depots—A country-specific distribution center and a concept worth considering in certain situations. For example, in countries that have proven difficult to ship to previously, or to which the transit time may be lengthy, such as some South American countries.
- Carriers and Procedures—There are a number of couriers available who will distribute CTS locally or globally, such as DHL, TNT, Marken, World Courier, and Air Express. These have their own procedures in handling and transportation.
- Cold Shipment Logistics—Distribution of CTS under refrigerated conditions.

Clinical trial materials should not experience variations in temperature, so great efforts are made to remove as many variables as possible to enable those involved to focus on the end points of the trial. Knowledge of storage stability and a product's ability to withstand temperature variances is only accumulated with time, and this expertise is not always easily available within pharmaceutical companies or, indeed, among pharmacists in hospitals (Evans 2000). A number of contractors are very experienced at distribution and know the importance of paying proper attention to the transportation and storage of temperature-sensitive products. They can advise and assist in improving the standards of the cold chain from manufacturer to patient or customer. However, there are only a few organizations that specialize in shipper validation, requesting dummy packs, physico-chemical data on all products to be shipped, details of destination(s), and the temperature at which the product(s) must be maintained. Also, hospitals generally may have good arrangements for storing temperature-sensitive products, but the standards in primary care and community pharmacy may vary enormously. Regulatory authorities periodically inspect and audit the premises they license (including licenses for manufacturing, assembly, and wholesaling), but they have no mandate to audit storage and distribution practices further down the line, particularly in retail and General Practitioner (GP) surgeries.

Increasingly, pharmaceutical and outsourcing companies are deploying Interactive Voice Response Systems (IVRS) and e-commerce systems. However, there are very few staff within the pharmaceutical industry who have experience with these systems as they apply to CTS and their warehousing and distribution. A contractor may be able to provide these services very effectively. Some contractors operate twenty-four hours a day, seven days a week; they have controlled temperature and humidity conditions and they also have validated processes for shipping and distribution.

Reconciliation and Destruction of CTS

In clinical trials, all medication for each patient has to be accounted for and returned to the sponsor or destroyed. Accounting is quite easy to do with high-technology tracking systems. However, some situations may be particularly demanding. For example, a pharmaceutical company may wish to conduct clinical trials with medicines required for tropical diseases, such as malaria, tuberculosis, or even HIV/AIDS in developing countries. Here, the tracking systems are extremely difficult to use, especially if a clinical trial takes place in a remote African village with no proper transport systems, or an Indian village that happens to be swamped with monsoon floodwater. Some companies have developed systems where all the unused supplies are returned to a central point for destruction. Local contractors can destroy these supplies, since one is only required to account for ensuring that these supplies are destroyed.

Process Improvement or Cycle-Time Reduction

In the conduct of clinical trials, there is often a huge waste of supplies. On the face of it, this may not appear to be a lot in terms of manufacturing costs. However, when all the different costs involved are added up, it does amount to a reduction of value to all concerned. These costs include: the warehouse space at the pharmaceutical company or the contractor, the shipment costs, the documentation required for each batch, the additional manpower required, the additional logistics, and the pharmacy time and space for CTS from a number of clinical trials from different pharmaceutical companies. The costs multiply with the number of trials being conducted. Therefore, in trying to reduce cycle times, elimination of waste is certainly important.

The Internet has created new opportunities for CTS service companies. First, it provides a means of providing information about the services they offer, and, more recently, a means of transacting business with customers online. Second, particular investigator sites can use it to order clinical trial material from a central point, hence avoiding stock buildup and the waste of any surplus CTS material. Some companies allow clients to review their inventory and shipping status of materials and packaged product via the Web, thus allowing clients to manage their development program more effectively. The use of IVRS in clinical trial management may help to significantly reduce surplus and waste, while providing savings in material and associated labor costs. There is a further benefit in flexibility and control; for example, it becomes easier to change the number of centers in a study or to use just-in-time techniques such as labeling and preparation of patient kits at the moment of dispatch (Noblet 2001).

Quality

Quality is the top priority; thus controls and checks have to be very thorough. One wrong pack in the clinical trial can put the entire trial material in doubt. There are few worse consequences for the actual clinical trial, which has taken a long time to set up and organize, and which is costly in terms of both time and money.

At present, medicinal products intended for research and development trials are not subject to European Community legislation governing either marketing or manufacturing. However, Member States may require compliance with the GMP principles during the manufacture of products intended for use in clinical trials. This is discussed in detail in Annex 13 of Manufacture of Investigational Medicinal Products (1997).

The European Union (EU) Clinical Trials Directive of 2001 has recently been approved. This directive must be incorporated into Member States' law by the end of 2004. At that date, all commercial clinical trials will have to be conducted using supplies that were manufactured in accordance with GMP.

THE MARKET FOR OUTSOURCING CLINICAL TRIAL SUPPLIES MANUFACTURING AND PACKING

At Technomark Consulting Services, we carried out the most recent survey of the CTS outsourcing market in 2001. The market survey covered Western and Eastern Europe and North America. We analyzed information for 189 contract organizations in Europe and 55 in North America that fit within our definition of a contract Clinical Trial Supplies Organization (CTSO). Some of the contract organizations have sites or plants in different countries; we counted each plant as a separate CTSO for this survey.

WHAT IS A CTSO?

We defined a CTSO according to whether it produces and/or packs pharmaceutical products for clinical trial purposes. The products may include the production of matching placebos and double dummy supplies. We excluded organizations that only offer labeling, printing, management, and/or distribution of supplies, even though these services represent an integral part of the CTS business. For this survey, we also excluded contractors or suppliers (primary manufacturers) of raw materials or Active Pharmaceutical Ingredients (APIs) and intermediates. A CTSO may be a contract packager and/or manufacturer (CPM) of pharmaceuticals or a contract research organization (CRO) as defined and described in the *Technomark Registers* (2001a; 2001b; 2001c). It may of course be solely a CTSO company.

Location

The number of CTSOs in each country is shown in Table 9.1. The UK has the largest number of CTSOs. Geographically, they are fairly evenly dispersed throughout the country, but there seems to be a concentration in the Southeast, the Northwest (around Manchester), and Wales. The distribution pattern may reflect the availability of regional grants. On the European mainland, about two-thirds of the CTSOs are found in Germany, Switzerland, and France. In North America, over 75 percent of the CTSOs are located along the Eastern Seaboard.

Classification of CTSOs

We grouped the known 244 CTSOs into three broad categories as follows.

Table 9.1   Location of Contract Organizations Providing CTS Services

Europe			
Austria	1	Ireland	6
Belgium	16	Italy	11
Eastern Europe	1	Netherlands	7
France	21	Scandinavia	12
Germany	39	Switzerland	16
Iberia (Spain/Portugal)	4	United Kingdom	56

North America			
United States	48	Canada	7

(Technomark market research data. Printed with permission from Technomark Consulting Services Ltd.)

Category 1—Contract organizations offering primarily CTS services

In this group, we include all CPMs or CROs that derive 50 percent or more of their business from the production and/or packaging of pharmaceutical products for clinical trial purposes.

Category 2—CPMs

This group includes CPMs that are mainly large, medium, or small production and/or packaging organizations whose main business is to manufacture pharmaceutical products under contract. They may also provide other services, such as formulation development, analytical testing, and so on, but CTS services account for between 1 and 50 percent of their overall business.

Category 3—CROs

This group includes CROs whose main business is to provide clinical research activities. They may also provide other services, such as formulation development and pilot-scale production, but CTS services account for between 1 and 50 percent of their overall business.

Figure 9.1 shows the CTSOs by the different categories for Europe and North America.

In recent years, many more CPMs have started providing CTS manufacturing or packing services, since they already have manufacturers or assembly licenses, but only a few of these organizations have dedicated CTS service units.

GROWING THE CTS BUSINESS

Figure 9.2 shows the number of CPMs/CROs providing CTS services in Europe according to their date of formation during 5-year intervals over the last 40 years (data includes 175 organizations). During the last 20 years, the number of CTSOs increased progressively at the rate of about 25 in each 5-year interval.

Figure 9.1   Contract organizations providing CTS according to their categories

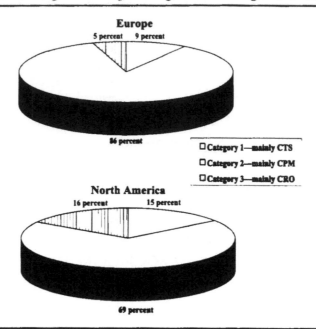

Europe

5 percent   9 percent

86 percent

☐ Category 1—mainly CTS
☐ Category 2—mainly CPM
☐ Category 3—mainly CRO

North America

16 percent   15 percent

69 percent

(Technomark market research data. Printed with permission from Technomark Consulting Services Ltd.)

Although these CPMs or CROs were formed in the years shown, they may have started to provide CTS services somewhat later. For example, only thirty out of the eighty-five CPMs/CROs that were formed before 1985 offered CTS services by that

Figure 9.2   Growth in the CTS business in Europe

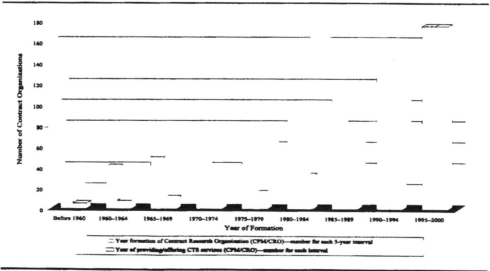

(Technomark market research data. Printed with permission from Technomark Consulting Services Ltd.)

year. Most of them did not begin to provide CTS services until the late 1980s or '90s. In North America, the contract market, especially the CPMs, is much younger than the European market, hence most of the CPMs started providing CTS services at or around the time when they were formed. Even though there have been some consolidation and mergers in the contract industry, growth of the CTS business continues despite there being no new companies formed in the last year or so.

Size of Market in Terms of Annual Revenue of CTSOs

We estimate the size of the market for contract clinical trial services to be approximately $400–500 million U.S. Although small in comparison with the market for outsourcing of clinical research, the clinical trial services market is dominated by a few players. Ten companies have, in total, 50 percent of the market. The CTS service sector has dramatically increased in size in the last ten years because it is potentially a high added-value business. It should not be regarded as a commodity business, in contrast to much of the secondary manufacturing business.

Size of CTSOs According to Number of Employees

The number of employees working in a CTSO can vary from one to another. In general, those CPMs or CROs that have dedicated or specialized CTS units deploy between 15 and 60 employees, depending on size. At the top end, two large CPMs stand out with 150 employees in their CTS unit, accounting for about 10 percent of the total staff employed by the companies. Otherwise, CPMs whose main business is in production and/or packaging of pharmaceuticals usually have fewer than 15 employees in the CTS unit. CTS unit staff need special training and usually include a registered pharmacist. Some CROs whose main business is clinical research usually deploy Clinical Research Associates (CRAs) and other staff who are also trained to work in the CTS unit.

Size of CTSOs According to Facilities and Global Scope

In the last three years, the largest CTSOs (fewer than ten) have expanded their facilities, which are in excess of ten thousand square meters. Otherwise, the average size of CTS facilities for packaging, manufacturing, and distribution is about fifteen hundred square meters. The major CTSOs have facilities in both North America and Europe; some of these have multiple European facilities. A few of these have or are planning to open up operations in Asia-Pacific regions, especially because of the potential increase in the number of patients available to enter into clinical trials.

CTS Services

There are 160 CTSOs in Europe and 45 CTSOs in North America (representing 84 percent and 81 percent of all CTSOs in Europe and North America, respectively) who provide manufacturing services (i.e., tablets, capsules, powders, liquids in syringes, and so on) with or without packing and/or labeling, randomization, and/or distribution (see Figure 9.3 and Figure 9.4). About 55 CTSOs in Europe and 13 CTSOs in North America (representing 29 percent and 23 percent of all the CTSOs in Europe and

Figure 9.3   European CTSOs by CTS services (n = 189)

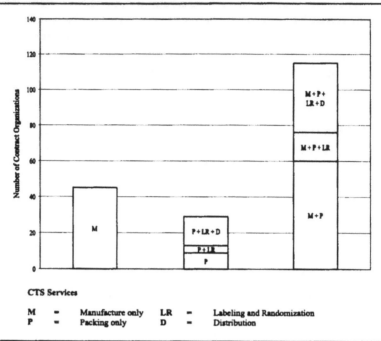

CTS Services

M	=	Manufacture only	LR	=	Labeling and Randomization
P	=	Packing only	D	=	Distribution

(Technomark market research data. Printed with permission from Technomark Consulting Services Ltd.)

North America, respectively) provide manufacturing and/or packing services with labeling, randomization, and distribution services (dedicated CTS units).

PHARMACEUTICAL COMPANIES' OPINION SURVEY

In a recent survey of thirty large and medium-sized pharmaceutical companies in Europe (UK, France, Germany, Italy, Switzerland, and Scandinavia) and North America, we found about 78 percent reporting that they contracted out their production and packaging of CTS. The remaining 22 percent said they had their own in-house facilities for preparing CTS. All thirty companies said they anticipated that their outsourcing needs would increase over the next two years (by about 20 percent).

The individual responsible for making the decision to outsource CTS production or packaging is most usually a CTS manager, clinical trials research manager, director of operations, or the medical director. However, there may be collaboration within a group involving these personnel who may well also co-opt the Quality Assurance (QA) manager and contracts manager. The person responsible for choosing a CTSO is usually the CTS manager or clinical trials research manager. The CTS manager or clinical trials research manager may liaise with the director of operations, contracts manager, QA manager, regulatory affairs manager, or formulation manager before making a decision about whether to choose a contract CTSO.

Figure 9.4   North American CTSOs by CTS services

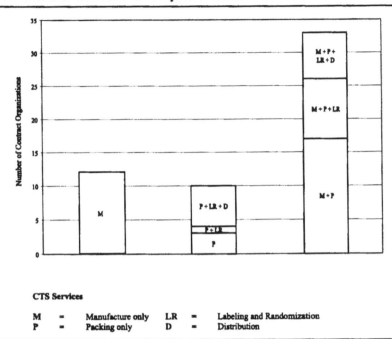

(Technomark market research data. Printed with permission from Technomark Consulting Services Ltd.)

Why Outsource Clinical Trial Supplies?

In our survey, the main reasons given for contracting out CTS production and packaging were that internal resources were overly stretched, the need for specialized packaging or a special formulation or technology, or the lack of special expertise or facilities in house.

Pharmaceutical companies have been finding it more difficult to keep pace with the increasing demands of clinical research as pipelines are being filled and where trials are becoming larger (with more patients) and more complex (for example, involving several dosing regimes). Also, these companies increasingly find they do not have the expertise, staff, or facilities necessary to produce all of their clinical trials supplies effectively. Sometimes pharmaceutical companies decide to no longer devote resources to set up in-house specialized capabilities or technologies for a single specific trial, especially when the risk of the product failing is quite high. This has led to an increasing demand for CTSOs that have the expertise, capabilities, and/or the specialized equipment, which they can use for several clients. In addition, CTS service providers are reputed to be cost-effective and respond quickly to requests. In the course of the survey, a medical director of an international company, although not typical, commented that ". . . our in-house clinical trial supplies are a shambles. One unit services clinical trial supplies for our European subsidiaries, centrally. Trial medication has to be requested formally. We work to tight schedules, but when the study is due to start, the trial medication isn't ready. Delays like that cost the company money. A CTSO would not let us down, if it wanted to stay in business. . . ."

Criteria for Choosing a CTSO

The most important criterion identified for choosing a contract CTS unit was "experience of working to GMP," followed, in decreasing order of importance, by "good quality control procedures in place," "availability to meet promised timelines," "ability to meet short lead times," and "ability to prevent/resolve problems."

Other criteria, such as "state-of-the-art facilities," "experience and specialized knowledge," and "purpose designed facilities," were important nonetheless.

CTS Services Availability

The most important service sought from a CTSO was labeling, followed by blister packing and drug accountability management (including returned product). Other services considered as important were randomization software, comparator medication supply, carding, distribution management, and bottle filling.

FORECASTING CTS BUSINESS—THE CHALLENGE

In the next few years, we anticipate that more documentation will be required for studies to meet Good Clinical Practice (GCP) guidelines. In order to obtain clinical trial data appropriate for meaningful statistical analyses, large numbers of patients need to be recruited. Furthermore, as the number of studies is likely to increase, the CTS business is projected to grow. CPMs or CROs need trained staff to deal with queries as well as to discuss the particular requirements for clinical trials with their clients' medical department personnel. Cost-cutting measures by major pharmaceutical companies aimed at downsizing their in-house facilities are making the economics of outsourcing clinical trial supplies look attractive. As pressure mounts up on a client's in-house resources, struggling to cope with the demand, the volume of CTS business contracted out is more than likely to increase. This is entirely consistent with the trend we found (see Figure 9.2).

We have identified a greater demand for more contract houses with facilities to GMP and U.S. Food and Drug Administration standards. We anticipate a greater demand for equipment and robotics in order to increase the throughput of CTS production, rather than trying to meet the demand merely by recruiting additional, costly staff. CTSOs will need to gear themselves to deal with the more complex packaging requirements for large, double-blind studies with improvements in packaging for patient compliance. A move to harmonizing packaging and labeling requirements for large, international, multicenter studies may pose another challenge for CTSOs in the near future.

Another factor that could contribute to an expansion of CTS business is the prospect of more clinical trials being conducted in Eastern European countries, Asia-Pacific countries (especially India and China), and South American countries. Whether that translates into CTS services being provided by CPMs or CROs located in those countries or new opportunities for Western European CTSOs would depend on the future path of European and North American political developments. These developments have implications for medicine testing legislation according to a common framework. The challenge facing CTSOs will be to successfully consolidate their existing business while exploring gaps in the emerging markets with the only certainty being that the European and North American political and economic climate will remain uncertain.

As biotechnology and emerging niche players bring their new products into development, most of them will have made a decision not to invest in setting up CTS capabilities internally. Thus, they will outsource to multiservice CROs or to specialist CTSOs. It is obvious that such a decision would be an integral part of the strategy of the virtual companies that have begun to appear in Europe and the United States.

Following the sequence and assembly of the human genome, the identification of the sequence of transcribed genes, the identification of genetic polymorphism in human populations, and the predicted development of individual therapies, it is generally anticipated that the absolute size of clinical trials for biotech products will decrease as smaller, more precisely defined patient populations are studied. However, the number of such studies is likely to increase substantially, since more trials in support of more labels will be undertaken. Thus the number of patients in studies will probably rise only slowly.

This scenario is a positive one for companies that can demonstrate flexibility and rapid customer reaction. The pressure on pricing is likely to be reflected in a higher unit price. However, patient kits are also likely to become more sophisticated and, thus, will justify the added value of this business. Further changes in this sector will probably entail additional investment in operators with packaging technology skills, as well as personnel with the requisite consumer marketing skills. As a result of these drivers, we expect a market growth of 10 to 15 percent in the outsourcing of CTS services in the next few years.

REFERENCES

Carmody, L. 2000. Global distribution of clinical trial supplies. *Pharmaceutical Manufacturing and Packing Sourcer* (June): 31.

Evans, J. M. 2000. Temperature-controlled distribution for clinical trials supplies. *Pharmaceutical Manufacturing and Packing Sourcer* (June): 38.

Manufacture of investigational medicinal products. 1997. Rules and Guidance for Pharmaceutical Manufacturers and Distributors. Annex 13: 159.

Noblet, M. 2001. Is there a revolution going on in clinical trial materials supply? *Pharmaceutical Manufacturing and Packing Sourcer* (March): 59.

The Technomark Register: European Contract Research Organisations. 2001a. Vol. 1. London: Technomark Consulting Services Ltd.

The Technomark Register: Contract Packagers and Manufacturers—Europe. 2001b. Vol. 3. London: Technomark Consulting Services Ltd.

The Technomark Register: Contract Research Organisations. 2001c. Vol. 4. London: Technomark Consulting Services Ltd.

21 Code of Federal Regulations 312, Subpart D. April 1999.

Ann Speaight

Contracting Out Laboratory Analysis

Most protocols for clinical trials involve laboratory analysis, either to ensure the safety of the subject, whether volunteer or patient, or to assess the behavior of the drug, or both. Few sponsor companies have the facilities to carry out the analysis (particularly the safety testing) in-house, so finding a lab to do the work is a regular requirement. In spite of the long history of contracting out lab work, there is surprisingly little structure in the way sponsors approach the task of appointing a lab.

The options and requirements in contracting out are different for each phase of drug development, but the choice of lab can have a marked effect on the study, not just in providing the analytical service, but in the quality of the data eventually submitted to data management for integration into the study database. It is important to bear in mind that laboratory data are not an end in themselves, but part of the body of data used to support a drug application. They can represent as much as 80 percent of all data generated in a study, so their ultimate use as part of the study database should be kept in mind. The focus of the investigator and the monitor is on the lab results; their interest is in ensuring the safety of the patient, and rightly so. However, if the data from a patient cannot be used in the database because of incorrect or incomplete demographic identification, one must question the ethics of dosing the patient in the first place.

Choosing a Laboratory

Before choosing and then assessing the laboratory that will perform the work in the study, there are some general factors that need to be considered.

Effect of the Protocol

In some studies, the study protocol defines the choice of which laboratory will carry out the analysis.

- If the study involves intensive care patients, there is little choice but to use the hospital lab, since the results are needed, often on an hourly basis, to treat the seriously ill patient.
- In Phase I studies, the lab must be close enough to the unit to ensure that results are available in time to meet the study's dosing points.
- In pediatric studies, the volume of blood needed often dictates the use of the hospital lab, where micro methods are available for analysis and appropriate reference ranges are used.

Using a Local Laboratory

A local laboratory is defined as a lab that serves the local hospital and/or community and that the investigator uses for nontrial patients.

There are many *advantages* to the site in using a local laboratory:

- The investigator has no problem with reference ranges; they are the same whether the patient is in a study or not.
- There is no problem with getting and storing blood collection supplies—they come in small batches from the local lab on a regular basis, as for nontrial patients.
- No problem exists with language or time differences in communications between lab and site.
- There is no problem with getting samples to the lab and no need to deal with couriers.
- In France, in particular, it is much easier to use local labs. This is because French doctors do not take blood and, unless the site is situated in a clinic, a phlebotomy service must be arranged to collect blood from the patients if the sponsor chooses to use a central lab.

The *disadvantages* of using a local lab are incurred mainly by the sponsor.

- Each center has its own local lab, which may not meet the required standard.
- It is difficult for sponsor data management to merge data because each lab has its own set of reference ranges and units.
- Resolving queries on lab data is difficult when there are so many labs involved.
- Results usually have to be transcribed onto case report forms (CRFs) by site personnel, because local labs use full patient IDs and not initials. This transcription can lead to errors.
- There is no possibility of direct electronic transfer of data to the sponsor database.

Using a Central Laboratory

A central lab is defined as one that analyzes all the samples in a study regardless of the location of the centers, or perhaps all the samples from centers in one country. The *advantages* in using a central lab are:

- One set of reference ranges and one standard of quality exist for all centers.
- Having one point of contact simplifies query resolution for the monitor and data managers.
- The sponsor project manager can track recruitment by center.
- The sponsor can audit the lab.
- Data can be directly transferred to the sponsor database.

There are also *disadvantages* in using a central lab, including:

- More work for the site, particularly in getting samples to the central lab, which may be in another country. Dealing with couriers causes more problems than any other issue in the use of central labs (IACL 1999).
- Communication between site and lab can be difficult because of language and time differences.
- The financial stability of central labs has become an issue. If the lab fails or is taken over, all the eggs are in one basket, so to speak.

Options with Central Laboratories

It is important to know if a central lab only analyzes samples from clinical trials, or if it offers a general clinical laboratory service with clinical trials as part of its portfolio. There are *advantages* and *disadvantages* in each case.

Clinical trials–only laboratory

- Theoretically, this is the best option because the lab staff understands what trials are about.
- The computer system is equipped already to deal with studies, visits, flagging protocol-specific results, and so on.
- There is a system for providing site support, trial-specific supplies, phoned results, and so on.
- One disadvantage is that the lab may not be able to perform all the non-routine tests on-site because these tests require a greater number of samples to perform economically. Three hundred Phase III clinical trials produce about 150 patient samples each day, and, often, studies only require the nonroutine tests on screening. These tests are often subcontracted, with a consequent effect on the turnaround time.

Mixed workload laboratory

- High throughput of samples—perhaps more than two thousand patients per day. Nonroutine and esoteric tests can be run on a daily basis.
- Cytology, histology, and microbiology are available on-site.

- The disadvantage is that clinical trial samples are usually a small percentage of the lab's total workload and do not get special analytical treatment because patients in trials are not perceived as ill.
- The lab's computer system is not always set up to link visits or verify demographics on patients in trials. Rather, these computers are geared toward acute patients and tracking systems that use full names and hospital or social security numbers rather than initials and numbers allocated in the study.

Virtual Laboratory

The concept of virtual laboratories is gaining favor as more studies become global. It is difficult to convince a physician in Australia, for instance, to wait for the results of a simple test to come from the other side of the world, when it can be done locally. The idea of the virtual lab is being explored to overcome the problem of integrating the data from different laboratories (perhaps a series of central labs throughout the world). The principle is that:

- All labs taking part in the study are sent standard material by the coordinator to analyze for the study's biochemical parameters;
- A protocol defines how often and in what configuration these samples are to be analyzed;
- Results are returned to the coordinator; and
- Correlation factors are calculated so that results from all participating laboratories can be converted to the same base.

This is the principle on which the national and international Quality Assurance (QA) schemes for biochemistry are based and it enables these schemes to compare results from and judge the performance of labs using different methods and analyzers, regardless of their location. Moving data, particularly for routine safety tests, is infinitely cheaper and more efficient than moving samples. For some very esoteric tests, it may not be possible to use this system, but, often, these results are not needed immediately, allowing samples to be shipped to one location and assayed in batches rather than one at a time.

Hematology and microbiology

The problem of comparing results from different laboratories applies to biochemistry assays, where there is a wide range of methods used for the same analyte, and which can, therefore, produce a wide range of results and reference ranges for the same tests on the same sample.

Hematology, on the other hand, does not have this problem. Hemoglobin is measured using a simple colorimetric assay, and red and white cells are particle counts, so it is easy to get universal agreement on values. The question of the quality of results produced by a laboratory remains, but the problem of the variety of assays and, therefore, reference ranges associated with biochemistry does not exist in hematology.

Like hematology, microbiology does not have the problems associated with biochemistry because bacteria grow on the same media and behave in the same way wherever they are. There is sometimes a question of characterizing certain bacteria and, for this, cultures can be centralized, but it is rarely an immediate requirement.

ASSESSING A LABORATORY

Local Laboratories

If the investigator chooses to use local labs in the study, it may not be practical for the monitor to visit the lab, but some assessment should be made of the facility, perhaps using a questionnaire submitted through the investigator. Results from these labs are used to assess the well-being of the patients taking part in the study; it is important to have some information on their standards and performance. Completing the questionnaire could perhaps be made a condition of allowing the investigator to use his local lab rather than a central one.

The monitor should know how the lab will present results before analysis begins and he or she should check a sample report to ensure that *all* the protocol tests are being carried out. Some tests are no longer considered relevant in hospital labs, but they nonetheless continue to appear in lab test protocol listings. If the local lab is not doing all the tests in the protocol, it is *vital* to address this, preferably before the study begins, but at least with the first patient's results. It may be that an unperformed test is unimportant and the medical director responsible for the study can agree that it need not be measured at sites where such a test is problematic, but the test may be the study's efficacy marker. In this case, arrangements can be made to have the assay carried out elsewhere, but these arrangements must be made at once; if the omission is not discovered until the data go to data management and the samples have been discarded, it is too late.

Central Laboratory

If a central laboratory is selected to perform the work, the primary objective should be to make life as smooth and uncomplicated as possible for the site—a happy site means a better study.

To facilitate smooth site operation, the following aspects of the service need to be assessed:

- Quality program and performance in external schemes
  - laboratory procedures
- Prestudy activities
  - investigator meetings
  - investigator manuals
  - request forms
- Getting samples to the lab
  - sample collection systems
  - couriers
- Reports
- Communicating abnormal values to the site

- Project management and data queries
- Data transfer and other computer-related facilities

Quality program

Assessing the quality of a clinical laboratory's results is a very specialized job. Ideally, it should be carried out by the sponsor company's QA group, but where this is infeasible, the monitor may have to act in this capacity. A professional QA consultant who is familiar with the workings of clinical laboratories should be asked to list the documents the laboratory should provide to the monitor. The QA consultant should scrutinize these documents and then provide a report to the sponsor. The process should cover both the internal and external programs used by the lab to assess the quality of results for all parameters involved in the study.

If the laboratory contracts out certain tests in the study, it is important to know how the reference laboratory is assessed and how the quality of the results is monitored.

Laboratory procedures. In addition to the quality of the results, it is important to understand how the laboratory works. The sites will never see the lab; the monitor will be acting as the interface between the lab and the sites. If the monitor understands the lab processes, he or she will be better equipped to deal with the sites' problems in relation to the lab.

The monitor needs to know:

- How samples are received, how they are labeled, and what checks are made to ensure that the correct sample is being analyzed;
- What the computer system checks at the data entry stage and if details from previous visits have been checked against new samples;
- What action, if any, is taken if a discrepancy is observed;
- How the information about the tests to be performed at each visit gets to the analyzers—via a manual or an automated system;
- How long samples are kept after analysis and if an option exists to keep them longer or have additional tests performed, if required;
- Whether out-of-range values are repeated automatically and how the repeat level is established; and
- Turnaround time: If some of the tests (e.g., cervical smears) are defined as exclusion or inclusion criteria, whether the lab can provide results in the interval between screening and entry visit.

Prestudy activities

Investigator meetings. It is very useful to have laboratory representatives present their part of the study at the investigator meeting, particularly if the sample handling at the site is complicated, where serum may have to be aliquoted and frozen before transmission to the lab (this seems to be increasingly common as sponsors seek to answer as many questions as possible in one study). Another benefit of having lab representatives at the meeting is to allow the lab to establish an identity. Because the site must enter or exclude patients based on the lab results, and, perhaps, act on abnormal results appearing during the study, it is important that the site has some confidence in the lab's performance. Seeing someone from the lab helps to establish this credibility.

Investigator manual. This is an important source of information for the site during the study, and it is important that the instructions are clearly laid out. If the study involves countries where English is not the primary language, the instructions regarding handling samples and transmitting them to the lab should be graphic as well as textual. The investigators in these centers may speak and understand English, but it is often a nurse or phlebotomist who takes the blood and he or she may not be quite so fluent. This is particularly important if the study involves complicated sample preparation for specialized tests, since assays can be totally meaningless if the sample has not been correctly collected and stored.

Request form. The request form is the means of communication between the site and the lab regarding the identification of the patient, the visit, and the tests to be performed. It defines the way in which the data will be collected—the number of characters in the patient number, the site identification, and so on. It also defines *what* data will be collected, so it is essential that data management has input into the design to ensure that all the necessary information is collected. Serious delays in finalizing the database will occur if a vital piece of information has not been recorded on the request forms so that it is held on the lab computer linked to the results files. Ideally, data management should provide a data protocol identifying how data should be identified and specifying detailed file structure format for data transfer (see "Data transfer" below).

Another question that must be decided is whether the request form should be a single form or a duplicate form with one copy going to the lab and one remaining at the site. The advantage of the duplicate form is that the monitor can check at the site if a sample has been sent, and can chase the lab if there is no report at the site.

Getting samples to the lab

Once a sample is collected from the patient by the investigator or the research nurse, it must be transported to the laboratory. Some of the issues involved in transporting samples are discussed next.

Sample collection materials

- What blood collection systems can the laboratory supply?
- How are the packs supplied (as components or assembled by visit, for instance)?
- Is there an expiry date clearly visible on the outside of the pack?
- Are sites *automatically* restocked when they have used all their kits or the kits have expired? If not, how does restocking happen?
- If bar-coded labels are used on the packs, what is coded on the bar code (if there is any site-specific information in the bar code, kits from one site must *never* be moved to another site to cover inventory shortages—the monitor needs to be aware of this)?
- How much notice does the lab need to send supplies to the site?
- How are supplies delivered to the sites? Is it possible to trace the signature of the recipient (this is particularly important in a hospital site where the courier cannot always identify the exact location of a particular investigator)?

Courier services. Most laboratories use couriers to get samples to the laboratory, although in the UK, using the post is still feasible. Dealing with couriers invariably creates problems for a site, and busy sites may have several different courier companies collecting samples every day. The lab usually chooses the courier. However, if the site has problems, they invariably come back to the monitor, not the lab. For this reason, the monitor should know precisely what services the courier will provide and he or she should consult the lab to verify that any critical study-specific issues have been addressed satisfactorily. Questions that might be addressed include:

- Will the couriers provide preprinted consignment notes for each site?
- How many signatures do they need on the consignment note (some courier services need three, which is often difficult to arrange [IACL 1999])?
- Can they track a sample from the moment it is picked up from the site, or only after it reaches the depot/hub?
- Can they provide latest times for booking and latest times for collecting from sites and can they *guarantee* to stick by them?
- If the site has booked in time and the courier does not pick up, is there a fall-back arrangement (this is particularly important for Friday pickups)?
- How does the booking system work? Is it sufficient for the site to give the account and center numbers, or do they have to repeat the complete site details each time they call to book?
- How long, on average, does it take to get through to make a booking? This should be confirmed by the monitor personally. The most frequent complaint about couriers is how long it takes to book a collection; if a site is dealing with several studies, it could take several hours each day (IACL 1999).
- What happens on weekends? If the courier picks up on Friday, can they deliver on Saturday? What additional charges are there for this service?
- Can they arrange dry ice collections, locally or internationally, if required? Do they have facilities to top up the dry ice if the shipment is delayed?
- Do they have custom clearance arrangement at airports (if applicable)?
- Do they supply all necessary paperwork for compliance with customs regulations if the samples have to cross international boundaries?
- Do they supply packaging for samples that complies with the International Air Transport Association (IATA) regulations for transport of a) not known to be infectious samples, and b) infectious samples?
- Do they have trained personnel available to handle infectious samples overseas?

Another point to be aware of is that although the courier should leave a copy of the consignment note with the site, they often don't. If they don't, there is no proof that they ever picked up the sample and it is the site's word against theirs that a sample ever existed. It is vital that the site insists on a copy of the consignment note for their records. The lab should make it clear to the site, but the monitor should ensure that the site understands that, without a copy of the consignment note signed by the courier, there is no proof the sample was collected and no hope of tracing it if it doesn't reach the lab. The monitor should follow up by checking for copies on occasional audit visits.

Reports

The laboratory can make managing the study much easier for the site by presenting results in certain ways. It is useful to inquire whether the lab can provide the services listed next.

Creating the report

- Can they print reports on three-part No Carbon Required (NCR) forms? Using NCR, the investigator can create, with one signature, a copy for the monitor and one for data management.
- Can they print boxes against abnormal values to allow the investigator to insert a code denoting the severity of the out-of-range value?
- Can they highlight or identify the values that constitute an exclusion value?
- Can they apply algorithms based on previous results for some pre-identified parameter (this might be needed in special cases and many labs can actually do it)?

Transmitting reports

- Can they fax to the site and follow up with hard copy?
- Will they send an additional copy to the monitor by fax and/or post if required?
- Will they send a third copy to the project manager at the sponsor company if required?
- Will they accommodate any combination that suits the sponsor?

Communicating abnormal values to the site

While it is important that the site knows as quickly as possible if a patient has a life-threatening abnormality, it is equally important to make sure that the site is not phoned with out-of-range values of no clinical significance. It is a good deal easier for the lab to send a fax than to make a phone call, but there is no guarantee that the information will get to the appropriate person. Direct contact by phone is required to ensure the safety of the patient with a clinically significant abnormality, and the lab must record who took the message and when. This contact procedure is time-consuming, making it even more important that the values defined as requiring phone notification are set at a minimum, consistent with the patient's safety.

Defining abnormal values for phoning to site. The first scan of protocol tests should aim to eliminate those that can never be life-threatening on an immediate basis (e.g., cholesterol or triglycerides), then to define those where an immediate answer will not instigate immediate treatment (e.g., urea or creatinine), and finally to arrive at the core tests where immediate intervention could be needed. These would include potassium, blood sugar, white cell count, and perhaps calcium, depending on the drug. Enzymes are more difficult because many things can cause raised liver enzymes, none of which are life-threatening in the short term, but raised enzymes have implications as to the subject's continuation in the study until a cause has been found, drug related or otherwise.

The medical adviser in charge of the study should make all decisions on levels requiring phone notification. The investigators should be given details of these decisions for their input, but the final decision should be the medical adviser's. With careful and realistic assessment of alert values, the number of phone calls needed should be few and relevant for most studies, and the lab computer can trigger a message alerting the project manager should any of the defined values occur.

It is sometimes difficult to decide how abnormal an abnormal value is—10 percent change in one analyte may mean nothing, but for another it could be serious. A useful booklet is one published by BUPA Medical Research & Development that reports *all* the results of the common safety assessment tests for 76,000 subjects, not just the 95 percent of values normally used to establish the reference range for laboratory results (Whitehead et al. 1994). The subjects are listed by age and sex and are presented in a way that allows an answer to the question "How often does this happen?" This booklet is particularly useful regarding raised enzyme results because it makes it possible to assess whether a value 20 percent above the upper limit of normal is a frequent or rare occurrence in a particular age group.

Project management and data queries

Besides the primary objective of ensuring a patient's safety in the study, there is a secondary objective of ensuring that the data generated by the laboratory can be readily integrated into the study database. For this, clean data are important, both the demographic data and the results, and securing clean data is one of the main tasks of the study project manager at the laboratory. Each sample must be tracked. If the request form is incomplete or the data inconsistent with data from a previous visit, the project manager must see that the site is contacted by phone or fax to obtain the missing information or resolve the discrepancy. If this is done by phone, it should be confirmed by fax or letter, so that there is a full audit trail for the data.

The aim is to make sure that every report sent to the site is correct and complete. It may involve issuing an interim report to ensure the safety of the patient and to prevent delay of results, but the final report should not be issued for the investigator's signature until all queries have been resolved and the amendments have been made. This is to ensure that the data in the CRFs, the laboratory computer, and the sponsor's database are the same.

The project manager should check also that all results are complete before study closure and that any agreed-upon codes for standard events (e.g., invalid results, hemolyzed samples, and so on) are applied before the data are transmitted, whether by direct modem link or on disk. The project manager is responsible for ensuring that defined abnormal values are telephoned to the investigator site and for highlighting protocol violations relating to lab values.

The project manager is responsible also for generating header listings. These will list all the demographic data recorded by the laboratory relating to the samples received and should be sent to the monitor for checking at predetermined intervals. If any errors are detected, the monitor should return the listing with the discrepancies noted so that the lab database can be amended. Where patient numbers were not assigned until randomization, the project manager must see that the screening visit is identified for each patient who has entered the study. The project manager can identify the relevant samples from the lab database, and, having confirmed these with the monitor, can allocate the patient number to the screening visit.

Data transfer

One of the main advantages of using a central laboratory is that the data from the lab computer can be transferred directly to the sponsor system, thus avoiding labor-intensive data entry. It sounds very simple, but in fact there are many hurdles, mostly having to do with the way the lab results are presented. Lab reports are documents that inform the investigator about patients' status; they are not structured in the exacting and consistent manner in which a computer expects to see them. For data transfer to work, the data management groups at the lab and at the sponsor company must communicate, preferably before the study starts, and agree upon procedures for dealing with the anomalies that will arise (e.g., text, where a numerical value was expected; values which are reported as ++, when the database expects to see a number; and comments, such as insufficient samples, test invalid, and so on). This process requires a good deal of work, but because lab data can be as much as 70 percent of all the data generated in the study, it is well worth spending time sorting out details. Procedures must be established regarding file structure as well as the actual transfer of data (how often and what data-only complete records? only updates? the whole study to date each time? and so on). There should be a test transmission using dummy data before the study starts in order to identify any system problems. Constant contact between the lab and sponsor data management/information technology (IT) groups should be maintained when the first records in the study are transmitted to make sure that everything works.

Data monitor. The sponsor company's data management group frequently encounters difficulties in coming to grips with lab data. Although the capacity to obtain data directly from a central lab has existed for some years, many sponsor groups still manually input the data from the lab reports. Their argument is that the data have so many variations, it is easier, when they are keying in all the other patient data, to input the lab data in the same way. They are reluctant to get involved with the laboratory even when encouraged to do so by lab personnel. However, direct data transfer saves time. If time savings are important to a sponsor company, it would be useful for the sponsor to have a *liaison data monitor* to act as the interface between the lab and data management at the sponsor group. Ideally, this person should have a laboratory background, preferably in clinical biochemistry, because most of the values are from the biochemistry safety profile and are also the values likely to cause the most difficulty.

COSTS AND CONTRACTS

Quotations for Studies

It is unnecessary to send the complete protocol when asking a laboratory to provide a quotation for a study. For the purpose of preparing the quote, the lab only needs to know:

- The number of patients;
- The number and distribution of the sites—national and international;
- The tests required and any tests that may need special preparation at the site and/or special storage and transportation conditions. This should include assays that may need to be stored for later assay by the sponsor;

- The required visits. This is best done by sending a copy of the study flowchart; and
- The estimated timeline, including the expected commencement date of the study, the estimated length of the recruitment period, and some idea of the expected completion date. The lab will be aware that all of these dates are estimates only, but, because the lab needs to schedule its workload, it needs to judge the sample throughput (is it likely to be one hundred samples a week or one hundred a year, for example).

Therapeutic area. Unlike other contract services in clinical trials, experience in the relevant therapeutic area is not important for the laboratory because the same block of tests is usually carried out for monitoring the safety of the patients' vital organs irrespective of the drug.

Additional charges. It is important to stipulate that the lab must provide *all predictable charges in the quote.* One of the most common complaints about laboratories is that the study always costs more than the budget defined by the sponsor based on the original quote. Sometimes overbudget costs cannot be foreseen (for instance, more patients had to be recruited because the number of withdrawals was higher than predicted) but sometimes, these costs are ones the lab may not have mentioned (for instance, fax handling charges or a fee for delivering supplies to the site).

Comparing quotes. The problem with comparing bids is that each lab has a different way of presenting information. A possible solution is to have a standard request format for all quotations, where the sponsor spells out each item and the lab fills in its charge for that service—if there is a charge—next to that item. This format must allow space for the laboratory to spell out what it provides in addition to the services requested, and at what cost. Data management and project management may be add-on services, both of which could make a substantial difference to the speed with which the final data can be integrated into the company database, but are not essential to the running of the study, strictly speaking.

Contracts

The system of contracts varies greatly from sponsor to sponsor. Most labs have their own contract and some sponsors also have standard contracts. Frequently, laboratories will begin work on the study on the basis of a letter of intent; in fact, it is not unknown for the study to be completed before the sponsor's legal department has finalized the contract to their satisfaction.

There are several pivotal things that need to be specified in the contract, including the level of service in terms of turnaround time, price, and quality of results, and premature termination arrangements for each side. The lab needs to be able to safeguard itself too, because sometimes a sponsor will commit to a study, the lab will order kits and equipment, perhaps take on extra staff, and the study is either terminated early for some reason or never gets going because of change of sponsor policy. The contract must include a clause allowing the lab to recover expenses incurred in setting up the study and, in some cases, to be compensated for loss of earnings where specific time has been booked on special analyzers.

The most difficult thing for a sponsor is when a central lab suddenly goes out of business. No sponsor can contractually protect against this, since even if the lab hopes to continue trading, one of its creditors may decide to foreclose. Nonetheless, a sponsor must do all it can to safeguard the results and to make the receivers aware of the importance of samples that may continue to arrive at the lab. Results can be keyed in from hard copy in the investigator's files, but if electronic data transfer is in place, it emphasizes the importance of regular data transfer—another reason that sponsors should consider electronic data transfer. So far as future samples are concerned, when an alternative laboratory has been found, converting new results to the old database values should be possible using a version of the virtual lab. If the initial lab took part in an external QA scheme, the coordinators of the scheme will have data on the equipment and methods used and will have factors allowing conversion of data obtained on the equipment used into values that match those produced on the new laboratory's equipment. If the new laboratory takes part in the same scheme—and there are not very many schemes, so it is not unlikely—the task is even easier. The schemes do not routinely offer this service, but some are beginning to consider expanding into the clinical trials market and this is one of the services they could usefully offer.

CERTIFICATION OF LABORATORIES

No national or international agreement exists yet regarding the certification of clinical laboratories that carry out testing on patients in studies. Preclinical work involving laboratory testing of animal samples is regulated and inspected by GLP (Good Laboratory Practice) inspectors and certified by the appropriate authority, which varies from country to country. Some laboratories that analyze samples from patients come under the GLP umbrella because they happen to work on animal samples as well. Clinical labs that don't analyze animal samples don't have access to this certification; for those labs, lack of GLP certification is no reflection on their performance as a clinical lab.

Several countries have an internal scheme for auditing clinical laboratories but these schemes focus on the clinical treatment of the patient and their emphasis is different from that of GLP. An important part of GLP is an analysis of a lab's processes, standard operating procedures (SOPs), training records, equipment maintenance records, and so on, in addition to an assessment of quality. In the UK, the CPA (Clinical Pathology Accreditation) is the scheme covering most of the National Health Service (NHS) labs and some of the independent labs. France and many other European countries have quite comprehensive national schemes, and in the United States, the CAP (College of American Pathologists) scheme is widely recognized by American sponsors. Various suggestions for overcoming the difficulties of auditing and certifying clinical labs have been made, including a proposal to create a GLP/C (meaning clinical) certification that has not moved forward because it failed to satisfy some of the international bodies involved.

A 1998 survey by the International Quality Assessment Company, Madesco (Desmet 1998) asked labs (both those doing clinical trials exclusively and those with a

Table 10.1   Certification Schemes Joined by Laboratories

Certification	Clinical Trials–Only Labs (7 labs responding)	Mixed Workload Labs (15 labs responding)
ISO 9002	0/7	3/15
EN-45001	1/7	4/15
IATA DGR	6/7	11/15
Annex 11 GMP	1/7	0/15
CAP (outside U.S.)	2/7	3/15
CLIA (outside U.S.)	1/7	
CDC (outside U.S.)	1/7	

mixed workload) which certification, if any, they held. Table 10.1 gives the distribution of their answers.

Most labs working in clinical trials understand the principles of GLP, and most have SOPs and all the appropriate records needed to comply with these regulations. Many sponsors carry out their own audits of labs and some have a list of preferred labs approved by their QA departments as satisfying their requirements.

CONCLUSION

Working with a laboratory to carry out analysis in clinical trials is a complex task. But it is worth remembering that the laboratory is as anxious as the sponsor to have a study run smoothly. If a study is not going as planned, speak to the lab about it. Often the lab staff are the last people to find out about problems relating to the lab, because it is often easier to simply blame the lab rather than trying to solve the problem. However, given the percentage of the data the lab supplies, it is worth the investment of time and effort to make it work.

REFERENCES

Desmet, M. 1998. Survey of labs. International Quality Assessment Company, Madesco. International Madesco Lindestraat, 29-B-3570 Alken, Belgium.

International Association of Clinical Labs (IACL). 1999. Survey of Study Site Coordinators (SSCs) on dealing with central labs. February. Copies of survey available from Ann Speaight.

Whitehead, T. P. et al. 1994. Adult reference values. BUPA Medical Research & Development Ltd. Copies available from Ann Speaight.

Jacqui Spencer

Contracting Data Management and Statistical Services

Clinical trials data are an essential asset for any pharmaceutical company, one of the few assets that truly increases in value over time. This data must therefore be valued and managed as an asset. For this reason, clinical data management is not something a data entry clerk or a data manager does. Rather, this critical function is performed by professional data managers. Their skill set does not revolve around creating tables in a database, but in identifying, describing, and acquiring (with clinical research colleagues) the data that makes up a regulatory submission. Clinical data management is about bringing quality to data. It is the practice of identifying problem data, verifying the source of the data, and facilitating the process of continual data improvement. It ensures that the collection of clinical trial data is continually improving the methodology, phases, and activities of a clinical trial.

As indicated in previous chapters, the business drivers within the pharmaceutical industry are unlikely to change over the next decade. Time to market will still be viewed as a critical milestone that needs to be reduced as much as possible to minimize the time to peak sales and maximize the duration of patient life for marketed products. As a result of these considerations, Contract Research Organizations (CROs) specializing in clinical data management and statistics have become, in recent years, increasingly important to the pharmaceutical industry.

In the 1980s, pharmaceutical companies had a clear opinion about which data should be processed internally and which could be outsourced. At that time, if the choice was between a pivotal study with a new chemical entity and a Phase IV study for a line extension on a marketed product, the latter project would have been outsourced. Companies wanted to maintain total control of key data because they were concerned about both the attitude of regulatory authorities and the confidentiality of the data. CROs quickly appreciated the importance of client confidentiality. Also in the early 1990s, the work of the International Conference on Harmonization (ICH) in streamlining the regulatory process resulted in a change of pharmaceutical company attitude to one where it was clear that regulatory bodies had no concerns around CROs processing pivotal data.

In response to this change in attitude, CROs have expanded the range of services they offer, drawing on the vast experience of their employees, who have typically worked in a varied marketplace. One result of this variety is that CROs now often have a broader knowledge of new Information Technology (IT) solutions than the clients they serve and are keen to offer this as a service to those clients. CROs have to respond more quickly to the ever-changing world of clinical data management to stay ahead of competitors and meet the needs of clients in order to win new tenders and maintain their existing client base.

Using CROs is now a viable and often cost-effective solution. The number of European CROs offering data management services (see Figure 11.1) means not only that a pharmaceutical company must face the difficulty of selecting the CRO that best suits its needs, but also that it must compete with CROs to recruit and maintain experienced clinical data management staff.

This resource issue is further compounded by the fact that many professionals still see clinical data management as a stepping stone to clinical research rather than a long-term career in its own right.

This chapter will consider the position of pharmaceutical companies involved in the outsourcing of clinical data management (which, for the purposes of this chapter, shall include data management and statistics), briefly reviewing the history of outsourcing the function and the various options available. The selection of a CRO for clinical data management activities offers specific challenges and this chapter will review these and

Figure 11.1   European CROs offering clinical data management and statistical services

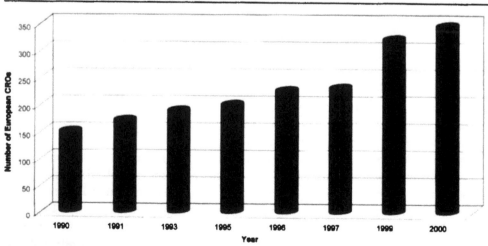

(Compiled from Technomark internal data. Printed with permission from Technomark Consulting Services Ltd.)

highlight the importance of the selection stage. Checklists will be provided to help identify the areas of most importance. Finally, the introduction of new technologies into the discipline will be discussed and an overview of future trends provided.

THE PHARMACEUTICAL COMPANIES' VIEWPOINT

Emerging markets (i.e., Latin America, East Asia, Eastern Europe, the Middle East, and parts of Africa) are already providing opportunities to accelerate patient recruitment into clinical trials through the availability of new, large, often "treatment naive" patient populations. Pharmaceutical companies will therefore be looking in these areas to ensure that the benefits of swift recruitment are not lost during the clinical data management process and will be looking to CROs to adopt data management methods that ensure continuing or even enhanced quality, but that decrease time to locked database.

In addition, clinical trials themselves are likely to be affected by regulatory and cultural changes. Governments will increase their interest in health economics and pharmacovigilance data and it is likely that they will become more interested in what might be termed "population medicine," which refers to small amounts of naturalistic data on large numbers of patients from across the world. This data could be collected using a simplistic form design (e.g., questionnaires) and therefore would be well-suited to specialist data management techniques such as electronic data capture (EDC). Pharmaceutical companies will be looking to CROs to provide a continuous, centralized data collection flow, not only up to product approval but also for the life cycle of the product.

In the mid-1990s, all submissions for regulatory approval from government authorities were paper-based (Hughes and Lumley 1999). Processing such submissions was time-consuming, resulting in costly delays for pharmaceutical companies. The United States Food and Drug Administration (FDA) was the first regulatory body to accept Computer-Assisted New Drug Applications (CANDAs). Such electronic submissions allowed the FDA to analyze the data more rapidly, thus markedly improving review times. The FDA announced that by 1996 all submissions of new drugs should be in the form of CANDAs. It later rescinded this decision, but today there is clearly a move toward CANDAs by the pharmaceutical industry. CROs have responded to this trend, seeing the provision of CANDA support as a key factor in their successful handling of outsourced clinical data management projects. Figure 11.2 shows the number of CROs providing CANDA support, broken down by country.

HISTORY OF OUTSOURCING CLINICAL DATA MANAGEMENT

In the 1980s, if a pharmaceutical company outsourced clinical data management to a CRO, that CRO was often faced with the challenge of transferring the data tape to the sponsor company to load into the in-house system. Today, the common practice is to provide the CRO with specifications to create a database like the sponsor's and to test data transfer periodically throughout the conduct of the trial. This approach is often flawed because of pharmaceutical company delays in providing database specifications.

Figure 11.2   CROs providing CANDA support (by country)

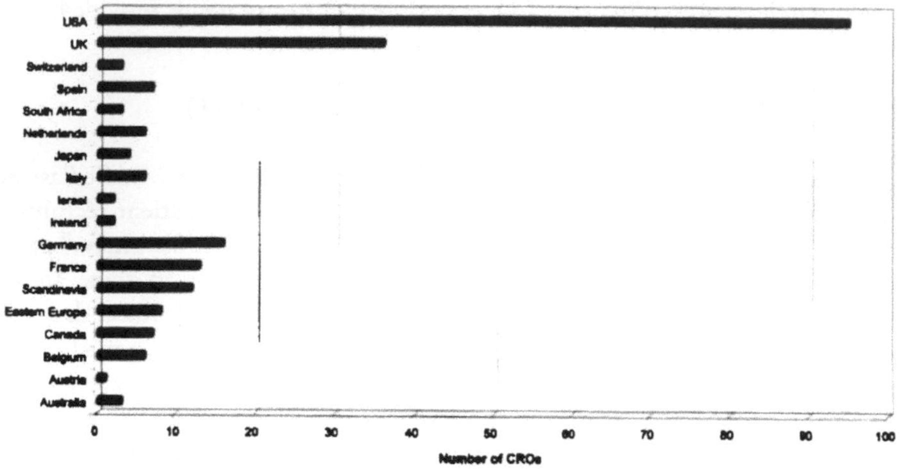

(*Technomark Register* 1999a; 1999b; 1999c; 1999d. Printed with permission from
Technomark Consulting Services Ltd.)

In the early 1990s, tactical outsourcing was made more attractive by the increase in
niche CROs specializing purely in clinical data management. When pharmaceutical
companies were faced with potential problems, they viewed the level of expertise and
the availability of support offered by these niche CROs as an attractive option. Indeed,
the first signs of *strategic* outsourcing within the drug development arena could be
attributed to niche clinical data management CROs, because a large number of pharmaceutical companies recognized these organizations as global data management centers and made a strategic decision to outsource all pivotal studies to a CRO. This was
seen as strategic outsourcing because it was a planned and intentional route, whereas
in the past, companies had outsourced studies for *tactical* reasons (i.e., because of staff
shortages or to increase specific expertise). The difference between tactical and strategic outsourcing is shown in Figure 11.3.

Today, pharmaceutical companies have readdressed the balance between tactical and
strategic outsourcing of clinical data management and adopted a more professional (and
successful) approach. The difficulty with outsourcing clinical data management tasks is
that it is still a relatively young discipline (the same cannot be said for statistical tasks,
but these are often seen as less problematic because of the industry's statistical approach
to standardization and the wide use of software like SAS®).

Clinical data management has really only been around for two decades.
Experienced professionals today often started at the data entry or graduate level and
learned their skills as their careers progressed. In the early years of clinical data management, overall project management was carried out by clinical research team members. To some degree, this is still true today.

Although the scene is changing, there is still a lack of experienced project managers
within clinical data management. While a clinical research associate's (CRA) function
typically includes financial and budgetary responsibilities for a study (or, indeed, a
number of studies), the statistician or clinical data manager function rarely includes
the financial aspects of project management. This lack of overall project management

Figure 11.3 Tactical versus strategic outsourcing

- Tactical
  - Manpower
  - Experience
  - Skills
  - Facilities
- Strategic
  - Competitive advantage
    - financial performance from increased long-term profitability
    - increased market share
    - new markets
    - faster time to market
    - increased marketability
  - Partnerships

experience, because it is an essential element for establishing a successful relationship between a pharmaceutical company and a CRO, directly affects the outsourcing of clinical data management.

To complete an outsourcing contract successfully today, a pharmaceutical company must possess both technical knowledge of clinical data management and project management skills. Many contracts fail either because the technical aspects of a contract are misunderstood (data standards, data format, encoding requirements, and so on) or because there has been poor communication and/or insufficient management of the CRO. As clinical data management professionals learn the additional skills required to work with CROs, and as it becomes standard practice for pharmaceutical companies to include their clinical data management staff at the start of a project, outsourcing becomes less problematic, resulting in more contracts being completed on time and within budget.

Outsourcing clinical data management also faces the additional challenge of an ever-changing IT environment. Consequently, how pharmaceutical companies outsource today will change dramatically over the next decade, as new technology becomes available. It is therefore important to continue to address the balance between technical and project management skills within data management departments.

TYPES OF CROs FOR CLINICAL DATA MANAGEMENT

The choice by a pharmaceutical company of using a niche CRO (e.g., biometrics or clinical monitoring) versus a global, full-service CRO for all program activities can be dictated by a variety of factors, ranging from an internal strategic directive to the need for additional expertise in a particular indication or therapeutic area. There is no doubt that the number of niche CROs offering only data management and statistical services is down in the last five years, but those that are still operating independently have often made a conscious decision not to expand into different sectors and to focus purely in the clinical data management marketplace (Hughes and Lumley 1999).

Whatever choice is made regarding the selection of CROs, it is imperative that the final objectives for the overall clinical program remain the driving force.

Four Scenarios for Using CROs

Table 11.1 shows examples of how clinical data management tasks could be managed when looking at four different scenarios of sponsor/CRO working relationships:

i.   Sponsor clinical monitoring with a *specialist biometrics CRO*
ii.   *Specialist CRO for clinical monitoring* with sponsor biometrics support
iii. *Different specialist CROs* for both biometrics and clinical monitoring
iv.   *A single (full-service) CRO* managing the complete program

Please note the emphasis has been placed on data management and statistical activities and, as such, each major task is represented. The other functions (e.g., clinical) have been grouped into main headings only (e.g., study design, monitoring) without specific task details.

Scenario i. *Sponsor clinical monitoring with a specialist biometrics CRO.* A sponsor's activities include the program design and initiation (with the CRO involved in a review, or secondary, role to establish complete understanding by both parties). Naturally, a sponsor then has total responsibility for the numerous clinical activities (summarized in Table 11.1 as monitoring and data collection). A sponsor's next primary role is the resolution of queries with the investigator. Following this, a sponsor assumes a review role (for some of the later tasks) until the completion of the final report.

Scenario ii. *Specialist CRO for clinical monitoring with sponsor biometrics support.* This obviously creates the highest potential workload for a sponsor's biometrics department. The time-consuming and complex activities involved with clinical monitoring become the primary responsibility of the specialist CRO; however there will still be an overlap with a sponsor's biometrics personnel during the query generation and resolution stages. As the CRO specializes in clinical monitoring, and their CRAs become familiar with the data, it is advantageous to include the CRAs in the table review steps and also as reviewers of the draft final report.

Scenario iii. *Different specialist CROs for both biometrics and clinical monitoring.* Using two CROs for different activities within a program requires continuous and dedicated coordination from a sponsor project management team. Naturally, an overlap develops for communication, roles and responsibilities, data ownership, and so on between a sponsor and each of the CROs, as well as between the CROs themselves. The possible difficulties can be balanced by the quality of the specialist manpower support provided and, consequently, the quality of the data.

Scenario iv. *A global (full-service) CRO managing the complete program.* This has become the most popular choice in recent years. The primary reasons to use a single CRO for all program activities are to achieve a continuous data flow through one system and to simplify communication. It is important to emphasize the fact that the program remains a sponsor's responsibility. Therefore, the assignment of a project management team and its involvement in guiding the program are extremely important.

Advantages and disadvantages

Scenario i. *Sponsor clinical monitoring with a specialist biometrics CRO.* With emphasis on quality in-house clinical design, setup, and execution of a study, it becomes easier to utilize a specialist biometrics CRO for the data entry activities, statistical analysis, or all

Table 11.1   Managing Clinical Data Management Tasks in Different Sponsor/CRO Relationships

Sponsor/CRO Working Relationship	Study Design	Study Initiation	Study Monitoring	Data Collection	Data Coding	Data Review	Data Entry	Data Verification	Data Validation	Internal Query Resolution	Correction Log Production	Investigator Query Resolution	Database Update	Database Soft Lock (Frozen)	Statistical Report Programming	Programming Review	Draft Table Generation	Draft Table Review	Database Unlock (Thaw)	Database Update	Database Hard Locked	Database Deblinded	Final Table Generation	Final Table Review	Analyses	Final Report
i. Sponsor clinical monitoring with a specialist biometrics CRO	S	S	S	S	A	A	A	A	A	A	A	S	A	A	A	A	A	A	A	A	A	A	A	A	A	A
	a	a	a	a	a	a				s	s	a						s							s	s
ii. Specialist CRO for clinical monitoring with sponsor biometrics support	S	B	B	B	S	S	S	S	S	S	S	B	S	S	S	S	S	S	S	S	S	S	S	S	S	S
	b	s	s	s		b	b			b	b	s						b				b			b	b
iii. Different specialist CROs for both biometrics and clinical monitoring	S	B	B	B	A	A	A	A	A	A	A	B	A	A	A	A	A	A	A	A	A	A	A	A	A	A
	a	a	a	a	b	b				b	b	a						b				b			b	b
	b	s	b	s	s													s							s	s
iv. A single (full-service) CRO managing the complete program	S	F	F	F	F	F	F	F	F	F	F	F	F	F	F	F	F	F	F	F	F	F	F	F	F	F
	f	s	s	s	s													s							s	s

S (or s) = Sponsor; A (or a) = Specialist biometrics CRO (include data management activities); B (or b) = Specialist clinical monitoring CRO; F (or f) = Single (full-service) CRO. Use of UPPER CASE letters indicates *primary responsibility*; use of lowercase letters indicates a suggested *review role*. For example: *Study design remains the primary responsibility for sponsor (S) for all scenarios, with a review role for the specialist clinical monitoring CRO (b) and the full-service CRO (f) in Scenarios ii and iv.*

activities relating to data management and biometrics. In order to maintain consistency within studies managed in house, the CRO is primarily used as an extension to the in-house data management and/or biometrics operations and therefore current in-house procedures relating to data management (i.e., query tracking and resolution) can be maintained. This option also allows contact with the clinical sites to remain within a sponsor's control.

It is important that an expert from within a sponsor's data management and/or biometrics group is part of the project management team to ensure that Standard Operating Procedures (SOPs) used by the CRO are in line with their own procedures and that all data standards, coding conventions and so on are followed.

Scenario ii. *Specialist CRO for clinical monitoring with sponsor biometrics support.* Using specialist clinical monitoring support is beneficial for a new clinical program within a new therapeutic area. Expert knowledge of the clinical data ensures quality at the start of the program.

If a CRO is used to perform the clinical monitoring activities because there are currently insufficient resources available, it is important to remember to look at the number of in-house personnel available to manage the activities both during the data collection period and during the postmonitoring phase (i.e., when the data arrives in-house.) There is little point in collecting the data quickly and efficiently by using specialist-monitoring personnel if internal resources cannot continuously manage the data flow.

As in Scenario i, the CRO must have an understanding of a sponsor's SOPs and have access to an assigned sponsor contact person dedicated to the project.

Scenario iii. *Different specialist CROs for both biometrics and clinical monitoring.* The quality of the data produced and the expertise of individuals managing each task are the major benefits in using this approach. But, a decision to run a clinical program with four different support groups participating (sponsor, investigators, CRO for clinical monitoring, and CRO for biometrics) should only be made if the overall project management approach is well-planned, with input from and consensus by all parties.

Communication and the overlap of responsibilities must be managed carefully with emphasis on written reports and frequent updates. A clear and well-defined contract and face-to-face meetings at important stages of the program are also required.

The relationships between the two CROs are obviously important. For example, a healthy rivalry may well benefit the management of the overall program, but, on the other hand, deadline difficulties may well result in conflicting opinions as to who is responsible for delays and so on.

Although the most complex of relationships, this system is used successfully by many pharmaceutical companies. However, it is recommended that a single manager perform the overall project management (across the four different groups). This requires support from additional personnel and, therefore, the impact on internal resources should be considered.

Scenario iv. *A global full-service CRO managing the complete program.* Project managers must be identified within the sponsor company and CRO and these two individuals should manage the overall execution and completion of the program (with larger support groups). Provided these project managers have the correct authorization levels, the overall project timings could be improved as the decision-making process has been simplified. Figure 11.4 summarizes the advantages and disadvantages of using a global CRO.

Figure 11.4   Advantages and disadvantages of using a global CRO

- Advantages:
  - Low management needs
  - Financially stable?
  - Resource rich
  - Experience rich?

- Disadvantages:
  - Variable quality from country to country and from function to function
  - Dependency on a single supplier
  - High prices
  - Lower priority compared with major clients or strategic allies

Program deadlines can be specified within the contract (e.g., the CRO is requested to provide a final report eight weeks after the last patient has completed the study, provided case report form [CRF] completion remains on schedule). Since patient recruitment is a known factor, both companies can monitor the status of the overall program.

Using a global CRO permits the data to flow through a single system (CRO dependent). This means that the management of the CRFs from the site through all steps indicated in Table 11.1 does not require the transfer of the data (or responsibilities) from one CRO to another and, therefore, the quality and consistency of the data are maintained.

Another consideration for the use of a single CRO is that many pharmaceutical companies with international offices experience some difficulties with the in-house management of multinational projects (e.g., language, local regulatory needs, local project needs, and so on). Since a global CRO tends to have ongoing experience with multinational projects for different sponsors, these problems have often already been addressed and, therefore, will have less effect on the management of multinational projects.

THE SELECTION OF CROs FOR CLINICAL DATA MANAGEMENT

The selection and management of CROs in general have been addressed in earlier chapters. This section relates purely to selection and management aspects that are unique to clinical data management. Consequently, general considerations, such as tender invitation, contract design, and financial management, are not discussed here. Also, more general criteria for selection, such as those shown in Figure 11.5, are not covered.

General Points for Consideration When Using CROs for Clinical Data Management

The factors to be addressed when outsourcing clinical data management activities often require specialist knowledge. It is therefore important that appropriate personnel are involved in the selection process and that the following points are considered:

Figure 11.5   Factors influencing the choice of a CRO

Experience*	Staff*
Expertise*	Personalities
Facilities	Training policies
Capacity*	Communications
Size	Quality*
Location	Reputation
Ownership	References
Financial stability*	Clients
Price	Accreditation
Chemistry*	Timeliness
* Key factors	

- Sponsor clinical data management resources should be identified to assist with project management.
- Consideration, at the start of the project, should be given to the future requirements for the data (e.g., ultimately the data from any clinical program will have to be integrated, either for a New Drug Application [NDA], an Integrated Safety Summary [ISS], or other regulatory activities, such as Product License Application [PLA] or Investigational New Drug [IND] and so on).
- The identification of the work to be contracted out is crucial. It is important to avoid, if at all possible, passing on partially completed projects (i.e., half-processed studies) or those that currently cause internal complications (the issues will be further complicated by the introduction of a third party).
- The advantages of using CROs may include the ability to process more data than a sponsor's in-house resources allow and, possibly, the completion of projects more quickly with less strain on internal resources.
- Disadvantages will include a certain loss of project control and, in new therapeutic/geographic areas, slower knowledge acquisition by internal clinical data management staff, but these can be minimized if a well-set-up and well-run contract is the focal point of the project.
- Quality Control (QC) checks should be part of the overall project plan and therefore should be managed and documented by the CRO completing the tasks, but these should be in line with any validation programs used within the sponsor company.
- Quality Assurance (QA) functions should remain the responsibility of the sponsor (or an additional specialized CRO). This means that QA should be undertaken by an independent resource qualified in clinical data management activities.
- Dictionary auto encoding (or any other coding conventions and/or systems) should remain consistent with in-house procedures/standards and therefore a QA check by the sponsor's clinical data management group is recommended (with particular emphasis on adverse event coding).

How to Select a CRO for Clinical Data Management

The successful selection of a CRO to undertake clinical data management activities depends heavily on the preparation of the tendor information provided to the CROs under consideration. This preparation can be time consuming and should be completed with input from "hands on" experts in the clinical data management area.

To obtain the information required to select the most appropriate CRO, one suggested approach would be to construct a prequalification questionnaire specific to clinical data management and provide this to the outsourcing project manager. This will ensure that all CROs approached will provide the same information and will facilitate the assessment of the questionnaire responses against clinical data management requirements.

Additional preparation prior to any assessment visit should include a thorough review of all company literature provided by the CRO, a review of additional publicly available information (e.g., CRO Web sites), and a second review of the questionnaire results provided. An expert in data management and statistics should be one of the sponsor assessors and this expert should arrange to meet the likely project manager, the head of data management, the head of statistics, the head of IT (or the CRO staff responsible for electronic data loading and/or transfer), and the person in charge of data entry/data coordination.

At the assessment visit, the CRO should be asked to provide the names of three relevant references who have worked specifically on clinical data management projects in the past. The sponsor expert in data management and statistics should also ensure that they review: appropriate Standard Operating Procedures (SOPs), clinical data management staff CVs and training records, the archive facility and archive procedures, clinical data management systems (in use), and the project management system. The areas for assessment below provide examples of clinical data management activities that could be addressed during the visit.

Areas for assessment

1. Clinical Data Management and Statistics
   1.1. Who heads the Clinical Data Management and Statistics department(s)? Obtain CV(s) if possible.
   1.2. How many other data management and statistical staff are there in each specific function? Which are full-time, part-time, and/or contract staff?

Category	Full-Time	Part-Time	Contract
Statisticians (more than five years' experience)	_____	_____	_____
Statisticians (less than five years' experience)	_____	_____	_____
Data managers	_____	_____	_____
Data-entry staff	_____	_____	_____
Programmers	_____	_____	_____
Other	_____	_____	_____
Other	_____	_____	_____
Other	_____	_____	_____

1.3. How does the data flow from the time a CRF arrives in house to the time of a database lock/analysis?

1.4. How are queries managed?

1.5. Who is responsible for tracking progress and with what?

1.6. What are the standard quality control checks used during the data management process and what is the CRO's standard error rates, and so on?

1.7. What computer systems/software are used:

    1.7.1. For clinical data management?

    1.7.2. For statistical programming/analysis?

    1.7.3. What version numbers are available?

    1.7.4. How are vendor products validated?

1.8. How is the Internet used for data management activities?

1.9. Who is responsible for (obtain CVs if possible) and how are the following carried out:

    1.9.1. Entering data (double/single manually, imaging and fax technology, and so on?

    1.9.2. Validating CRFs (manually/electronically)?

    1.9.3. Writing the validation plan?

    1.9.4. Writing validation programs?

    1.9.5. Writing the statistical plan?

    1.9.6. Writing statistical programs?

    1.9.7. Producing listings and tables?

    1.9.8. Writing the report?

1.10. What is the company's anticipated workload/capacity?

1.11. Who trains the staff and maintains training records?

1.12. How are difficulties managed with regard to:

    1.12.1. quality of data?

    1.12.2. timeliness of CRF/query flow?

2. Project Management

2.1. How does Project Management operate in the organization?

2.2. Who are project managers? What backgrounds do they have? Obtain CVs if possible.

2.3. How much responsibility does the project manager have (e.g., internal resources, client, budget, training, and so on)?

2.4. How many projects does the project manager have at any time?

3. SOPs

3.1. How are SOPs written? By whom? Who authorizes them?

3.2. How often are they reviewed?

3.3. Who distributes them and how?

3.4. How is SOP training provided?

3.5. What SOPs are available on data management and statistical processes?

4. Archiving and Storage Facilities

4.1. Where are CRFs and backup media stored and archived?

4.2. What fire precautions are there?

4.3. What disaster recovery options are available?

5. Overall Impression of Clinical Data Management Capabilities

Finally, if required, the responses to the prequalification questionnaire should be discussed and any outstanding questions resulting from the information provided prior to, or during, the assessment visit should be addressed. In order to achieve the best results from this final stage, it is advisable for the sponsor team visiting the CRO to include in the assessment agenda a time slot for the team to meet without CRO staff present. The final agenda item should be a meeting where all CRO and sponsor representatives gather to cover any outstanding items and close the visit.

Selecting the CRO(s)

In addition to the information provided in the initial CRO package and the findings from the assessment visit, it is important to include a weighted analysis that provides a numerical rating of the CROs from every member of the selection team. Although this rating should not be the basis on which final selection is made, it provides a structured and impartial element that should be considered with the other information available.

The project manager should design and set up a spreadsheet containing all of the criteria that are important for selection (see Table 11.2 for example criteria). Each of these criteria should be weighted in order of important (e.g., weighting of 100 for clinical data management expertise, weighting of 40 for the CRO presentation to the sponsor). These weightings should then be added as formulas in the spreadsheet.

Once the criteria and weighting have been completed, a paper copy relating to each CRO should be provided to each member of the selection team. The team should then score each of the CROs against each criterion and return responses to the

Table 11.2   Weighting Criteria for CRO Selection

Criteria	Weight
Staffing	60
Data Flow	60
Therapeutic Experience	80
CRO Presentation to Sponsor	40
Feel-Good Factor	70
Response to Questions	50
Technical Capabilities	80
Interpretation of Brief	50
Data Management Expertise	100
Statistical Expertise	90
Quality Assurance	50
Total	**730**

project manager. The responses should then be entered into the spreadsheet and the formula for the weighting will then provide total rating for each of the CROs.

The selection team should then examine the results. Does the team feel comfortable with the result? If not, reexamine the criteria weighting and adjust if absolutely necessary—do not ask the selection team to provide new scores as they will have been influenced by the preliminary result.

The final results from the weighted analysis should then be discussed in conjunction with all other CRO information and a selection decision reached.

Summary

The overall selection process is one that requires a great deal of sponsor preparation. Although weighted analysis techniques are not essential, they are often seen as a good basis for discussion and they eliminate the bias that can naturally intrude when a number of assessment visits are performed within a short period of time. Such techniques also allow for a more structured and standard approach, one that can be repeated easily. It should be noted, however, that the weighted analysis is only one element of the assessment process and it should only be used as a form of guidance, alongside people's impressions, cost and time estimates, and so on.

FUTURE TRENDS: INFORMATION TECHNOLOGY

In the past three years, the increasing number of companies offering Web-based solutions to clinical data management has given a clear indication that the future of clinical trial data management will be most affected by advances in information technology (IT). For example, as EDC utilizes many different aspects of IT, it has the potential to change dramatically. Sophisticated EDC capabilities could emerge as a key source of differentiation and competitive advantage for pharmaceutical companies and CROs over the next three to ten years (Mihkin 1998).

Today, companies and patients alike are defining the way the Internet is used and can be used for clinical trials. Pharmaceutical companies use it to communicate information about the diseases in which they specialized and to offer product-specific promotions. Patients are primarily interested in locating product information and details relating to their health or potential medical treatments including listings of ongoing clinical trials (Johnson and Wordell 1998; Pines 1998). Even though the Internet is currently more widely used in the United States than other countries, the benefits of rapid communication and data transfer are resulting in widespread interest in the use of the Internet as a centrally served Web-based application. See Figure 11.6 for the collection, management, and analysis of clinical trial data.

Early problems associated with using centrally served Web-based applications, such as slow bandwidth (influencing speed of use) and security of data, have largely been addressed by improved hardware and software capabilities (i.e., encryption) and the introduction of new regulatory guidelines like 21 Code of Federal Regulations Part 11 (1997). The explosion of digital technology is an indication of how communication networks will continue to change. Although not all countries used for clinical trials will have access to such networks by the end of this decade, alternative solutions will no

Figure 11.6   Centrally served Web-based application

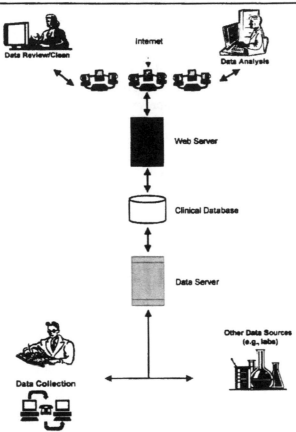

doubt be available (e.g., satellite communication networks). Technology will not be the only aspect of clinical data management that will change. The job functions and skill levels involved in the management and collection of clinical trial data will have to change in order to support this new environment and all pharmaceutical companies and CROs involved will have to adjust accordingly (Regalada 1998).

To differentiate themselves from competitors, or to provide the complete solution, a number of CROs have already moved toward providing technical solutions based on their increased IT knowledge base (Hughes and Lumley 1999). Although offering technology as a service will attract pharmaceutical companies that are reluctant to invest in IT solutions, CROs could miss out if the trend in pharmaceutical companies toward a single internal solution to all IT needs continues. Pharmaceutical companies may well look toward strategic technology mergers, acquisitions, and alliances that address all internal IT needs and fall outside of the capabilities of CROs whose IT services are secondary to their focus of supporting the outsourcing of clinical data management activities.

Pharmaceutical companies and CROs will need to adapt to the changing IT environment and, in particular, data management professionals will need to support the use of new data management methods. A centralized Web-based system that receives data from numerous sources could replace a central clinical database maintained by manual data entry. All third parties (e.g., central laboratories and CROs) would have direct access to such a system and the data would be transferred directly without the need for manual entry.

All trial documentation could be available on study-specific Web pages. Revisions to and distribution of documentation (e.g., new SOPs and protocol amendments) could be centralized and investigators could have access to up-to-date study information. Chat rooms could be available for site personnel to share their experiences, and e-mail links to study monitors would improve communication.

In addition, automatic online coding of trial data would improve the quality of the data collected, allowing the site to provide accurate data at the point of entry. Making automatic e-mail notification of serious adverse events part of the adverse event form available within the Web-based system would allow safety data to be more easily managed. Query generation, management, and resolution could be electronic, thus reducing the requirement for frequent site visits. The data would be cleaner than traditional data entry methods allow and, as a result, the database could be locked within hours of the final patient visit.

In the long term, there may be a limited need for data entry staff and the responsibilities of data management staff may shift toward project management. Depending on the structure of the pharmaceutical company or CRO and the skill sets of available personnel, it may be possible to combine the roles of the clinical data manager and the clinical research associate. However, it is likely that not all trials will be suited to the use of new technology. Some countries will take longer to equip themselves to use Internet technology effectively and some clinical studies using paper CRFs will be more effectively managed or more cost-effective. Therefore, pharmaceutical companies and data management staff in particular will need to be in a position to utilize the latest technology wherever possible, while remaining able to use traditional methods where appropriate and able to assess which is the most appropriate approach to take.

CONCLUSIONS

If contracting data management and statistical services will result in an increase in quality and a decrease in project timings, then it becomes an attractive option. However, to truly evaluate the benefits, there must be an internal benchmark to which contracting can be compared. Pharmaceutical companies must know what is realistically achievable, for what cost, and within what time frame.

The clinical data management staff within a CRO and pharmaceutical company are peers. All involved should have comparable technical knowledge and work as a partnership. They should play a part in the management of the project.

The difficulties that exist within a project managed internally (e.g., poor protocol design) will not be resolved by outsourcing. In fact, they will escalate. If existing processes fail, then repeating them will not increase the likelihood of success.

Clinical data management is a discipline that can never be static. If the people, projects, priorities, and technology can change at a moment's notice, then the emphasis must be on the processes and procedures. Training and development, communication and project management, combined with clear and accurate processes and procedures, offer the answer to success for both CROs and pharmaceutical companies alike.

Clinical data management professionals pride themselves on analysis of information and attention to detail. If these skills are utilized when contracting data management and statistical services, then success will be almost guaranteed.

REFERENCES

Code of Federal Regulations. 1997. Title 21. Part 11. Electronic records; Electronic signatures. Washington, D.C.: U.S. Government Printing Office.

Hughes, R. G. and C. E. Lumley. 1999. *Current strategies and future prospects in pharmaceutical outsourcing.* London: Technomark Consulting Services Ltd.

Johnson, S. T. and C. J. Wordell. 1998. *Drug Information Journal* 32: 547–554.

Mihkin, A. 1998. *Break-out opportunities in pharma IT in development.* Paper presented at DIA Annual Meeting 11 June.

Pines, W. L. 1998. *Drug Information Journal* 32: 227–281.

Regalada, A. 1998. Drug development II: Clinical data collection and management. *Start-Up.* Windhover Information Inc.

Technomark Register: European contract research organisations—clinical research. 1999a. Vol. 1. London: Technomark Consulting Services Ltd.

Technomark Register: European toxicology and analytical organisations. 1999b. Vol. 2. London: Technomark Consulting Services Ltd.

Technomark Register: Contract packers and manufacturers—Europe. 1999c. Vol. 3. London: Technomark Consulting Services Ltd.

Technomark Register: European contract research organisations—North America. 1999d. Vol. 1. London: Technomark Consulting Services Ltd.

Madeline J. Ducate

Global Clinical Trials: The Cultural Challenge

Over the past decade, global clinical trials are rapidly becoming more the norm than the exception in clinical drug development programs. Economic factors and global communication developments, combined with the International Conference on Harmonization (ICH) initiatives, have resulted in more and more pharmaceutical companies planning and executing clinical trials using a mix of locations to optimize their drug development timelines. Thus, global outsourcing implies seeking one or more outsource vendors who can supply services or data beyond your country's borders and usually on more than one continent. The multinational and Pan-European clinical trials take a backseat today to the complexity, planning, and execution needed when different time zones, different continents, and different cultures are combined. Ultimately, despite this mixing, homogeneous results must be obtained to answer the clinical objectives of the study.

Clinical development plans today typically take into account the marketing authorization requirements for more than one country in order to maximize the return on investment for a compound's research and development. The use of "foreign data" as pivotal and supportive data to a New Drug Application (NDA) or a Marketing Authorization Application (MAA) requires the assurance of common or uniform standards of quality and practices. ICH Good Clinical Practices (GCP) can provide some of these assurances, but in reality, it is the diligence of the research teams that ensures a global trial employing up to hundreds of sites on several continents will produce a meaningful database through which sound analyses and decisions can be made.

The intent of this chapter is to highlight some differences and some similarities in conducting trials in a wide range of different countries. It will also offer some practical experience in managing the differences, so that multiple studies in different countries can be analyzed singly and together, and global trials can produce uniform data for comparative analyses.

GLOBAL PROGRAMS VERSUS GLOBAL STUDIES

Global clinical development programs involve the combination of one or more clinical trials conducted in various parts of the world. The expectation is that the data are meaningful to, and can be applied to, the patient population wherein the drug will be marketed. More and more frequently, the clinical aspects of drug development programs are being designed globally, as full-scale programs, i.e., having a series of studies conducted in various parts of the world that can be used pivotally and supportively for multiple market authorizations.

On the other hand, a single study conducted globally is generally undertaken when the expected patient enrollment rate is low and the number of patients required is high. Such trials are sometimes referred to as megatrials and are often seen in mortality outcome designs, such as heart failure studies, stroke, severe sepsis, and so on. Table 12.1 summarizes the differences between a global program and a global study. Global studies are the most challenging type of clinical trial today. Global trial management can take on various shapes, and outsourcing requires experience and considerable forethought prior to contacting potential sites. Planned, global trials require adequate feasibility not only to justify the increased cost, but also to reasonably ensure completion on time and with suitable and comparable data.

Occasionally, we have seen that a clinical study that fails to meet intended timelines may start as a multicenter trial in more than one country and require additional sites/countries to supplement the current trial. Adding countries and changing sites after the original plan adds to the complexity of the trial with an exaggerated staggering of site initiations. It is important that such studies are handled carefully so that the patient population remains unaltered and the data remain of consistent quality. Figure 12.1 presents some basic questions that can be used to determine whether a global program or a global study is appropriate and warranted.

In both situations, some fundamental questions, such as those presented in Figure 12.2, should be asked very early in the decision-making process. Global outsourcing involves working with contractors who have experience in understanding what it takes

---

Table 12.1   A Global Program Versus a Global Study

Global Program	Global Study
• Development program involving multiple studies conducted in more than one country, generally with the intent to file multiple NDAs/MAAs	• A single study conducted in multiple countries crossing cultural, geographical, and time barriers, thus requiring extensive management
• Ultimately the goal for almost all drug development programs undertaken today	• Generally required for critical care studies, mortality endpoint trials, orphan indications, and complex disease entities

Figure 12.1   Is a global program or a global study appropriate and warranted?

Challenge the reason for global studies versus global programs:

1. Is patient recruitment the driving factor? Will it be difficult to recruit patients in a reasonable time period? Is the indication a rare disease? *Answer:* A global study may be appropriate.

or

2. Are marketing departments interested in having their country participate? Is it important to have a variety of opinion leaders interested in participating? Enrollment is not a concern. *Answer:* A global program encompassing a variety of studies from different countries is simpler and more cost-effective to conduct.

Figure 12.2   Questions to ask prior to undertaking multinational or global trials

- Is patient population for the trial clearly defined similarly in the various countries anticipated for inclusion? Are the treatments for this indication already on the market in these countries? Are they the same treatments? Are there nondrug treatments in these countries that may make a difference in the treatment groups?
- Is the active comparator on the market in the same formulation and strength as required by the protocol?
- Are there any regulatory issues concerning the manufacture or importing of not only the investigational product but also the comparator, including placebo? What is the composition of the placebo? Will there be any import concerns, courier challenges, and so on to meet?
- Are the countries/sites equipped to conduct evaluation in a uniform manner? Is the medical practice for these procedures the same or similar?
- Will local or central laboratories be used? If local, will the laboratory data be comparable for the study? If central, will sample transport be a problem in terms of timing, temperature conditions, and so on?
- Will there be competing trials in this clinical indication?

to conduct these trials and who have experience in understanding the need not only to know the answers today, but also to keep current with the ever-changing regulatory, medical, and clinical practices worldwide.

This chapter looks at various aspects of conducting a global study and is not meant to be inclusive of all considerations that should be made in the planning stages of such trials. Rather, it is intended to provide a *flavor* of some issues, potential problems, and differences that may be encountered. It places an emphasis on handling more complex and complicated trials, such as those conducted in Intensive Care Units (ICUs) or other institutional specialty units. Eight topics will be covered: enrollment considerations, patient follow-up/retention, medical practices and healthcare equipment, investigational fees, central randomization and medical hot lines, central and local laboratories, pharmacies, and importing investigational products, followed by a case study illustration.

ENROLLMENT CONSIDERATIONS

Experience with numerous global trials demonstrates that no two trials are alike, nor are they guaranteed to be predictable, since standards of practice and care vary over time. In addition to a dedicated and committed investigator, it is vastly important that the hospital management supports the investigator, especially with regard to the conduct of clinical research in an ICU or other specialty unit. The investigator should have a stable and, preferably, dedicated research team. Experience has shown that countries such as the United States, Canada, UK, Sweden, the Netherlands, Australia, and others typically have clinical research units in medical facilities that are supportive to investigational studies. An investigator's availability and flexibility are measured by his or her:

- Ability to devote time during the prestudy visit;
- Timely response to essential document collection;
- Involvement and follow-through with Institutional Review Board (IRB)/Independent Ethics Committee (IEC) activities; and
- Supportive and active contributions to ensure that his or her team is prepared, protocol-knowledgeable, and ready prior to the initiation visit.

In preparation of the planned initiation visit, sites should be asked to provide a three- to four-week retrospective review of potential patients that were in the unit in order to evaluate the screening activities that they will actually conduct once the investigational product is provided. During the initiation visit, the actual charts of potential patients should be reviewed thoroughly in terms of inclusion/exclusion, time to meet the criteria assessment window, source data management, consenting process, randomization process, and so on. By utilizing actual cases, the initiation visit shifts some of the "if," "when," and "how it can be done" discussion to "how it should have been handled in these cases." If the investigational staff is involved in another trial, these resources may be stretched too tightly in the department; therefore, it is best to insist that these sites either serve as backup or get scheduled to begin at a later date.

The study itself should not overlap another study in terms of patient population and definitions, and it is ethically important that these sites complete enrollment for the competing trial prior to initiation into the new study. Because of the lag time between prestudy and initiation visits, it is important to continually assess the staff availability before the investigational product is delivered to the unit.

Site Management Organizations (SMOs)—especially hospital and specialty SMOs—are emerging in Europe and other countries outside the United States. SMOs in the United States are also redefining themselves and their services, and should be considered along with other sites in global studies. Therapeutic medical organizations and investigator consortiums are also interested in participating in such trials and their medical consulting expertise is often used for committees associated with and required for these types of trials, such as Data Safety and Monitoring Boards (DSMBs) and Clinical Evaluation Committees (CECs). Institutional SMOs have emerged primarily in response to the lack of organization in hospital research offices, the need to consolidate resources, the formalization of IRB/IEC activities, ICH GCP requirements, and financial opportunities for the hospitals and medical centers. Contract Research Organizations (CROs) and global outsource vendors should have a specific Standard Operating Procedure (SOP) for evaluating SMOs in general and for determining appropriate working relationships with them. Hospital SMOs are located in the UK, the Netherlands, and Sweden and they are beginning to develop in Central Europe, Germany, India, and other countries. CROs should collect and maintain an SMO database and, more importantly, should independently evaluate SMOs for potential use in future studies, with an emphasis on their ability to contribute to global studies as well as single and multinational studies.

SMOs organized around primary care physicians and noninstitutional sites are also a source of standardized care and research-dedicated staff for outpatient global trials. These SMOs should be investigated and relationships with them should be nurtured in the same manner as described above.

It is essential that large trials with anticipated enrollment rates of less than one patient per month per site be closely managed from the start. One important metric to record and manage closely is the time from initiation visit (with the investigational product on site) to first patient enrolled. The longer the delay, the more hesitant the investigator and staff are to test the procedures needed to comply with the protocol. When the anticipated enrollment is low and the medical unit's workload is high, it is essential that screening logs and other data be sent on a regular basis from the sites to the Clinical Research Associates (CRAs). Calls to a medical support line or hotline from the sites can be used as a surrogate measure of real-site activities (even when enrollment is slow). The sponsor's study management team should develop a plan to measure how long it takes for a site to become active (enrolling patients). Table 12.2 illustrates the calculations of these enrollment rates. This plan would have a predefined time window prior to remedial steps being taken, including "PR" (public relations) visits, reinitiation, and so on. Once a center enrolls their first patient, it is important to conduct an interim visit to ensure that procedures are being followed, that the site is comfortable with the activities, and to encourage them to continue at a similar pace.

Table 12.2   Patient Enrollment per Site per Month

Patient Enrollment per Site per Month
Study Name/Study Drug
Protocol Number: 123456
As of 7 January 2001

Investigator (Site No.)	Initiation Visit Date	Date Drug on Site	No. of Patients Randomized	No. of Months Active	No. of Weeks Active	Average Enrollment Rate
Bxxx/Keidelberg (3401)	18 Dec 1999	18 Dec 1999	14	12.65	55.03	1.11
Cxxx/Geelong (3402)	14 Dec 1999	07 Dec 1999	11	13.05	56.60	0.85
Dxxx/Perth (3404)	11 Jan 2000	14 Jan 2000	2	11.78	51.17	0.17
Fxxx/Footscray (3405)	17 Nov 1999	07 Dec 1999	4	13.01	56.60	0.31
Mxxx/Kentworthville (3430)	22 Apr 2000	25 Apr 2000	0	8.39	36.60	0.00
Pxxx/Woodville (3406)	14 Oct 1999	18 Oct 1999	7	14.62	63.60	0.49
Rxxx/Woolongong (3409)	25 Mar 2000	29 Mar 2000	6	9.30	40.60	0.65
Txxx/Hobart (3407)	15 Oct 1999	26 Oct 1999	3	14.39	62.60	0.21
Total for Australia			**47**	**97.15**	**422.80**	**0.48**
Jxxx/Warsaw (2002)	29 Jul 1998	29 Jul 1998	15	29.30	127.46	0.51
Kxxx/Warsaw (2003)	19 Jun 1998	19 Jun 1998	11	30.62	133.17	0.36
Kxxx/Sosnoniec (2004)	25 Jun 1998	09 Jul 1998	20	29.97	130.46	0.67
Kxxx/Wroclaw (2005)	26 Jun 1998	06 Jul 1998	33	29.97	130.46	1.10
Sxxx/Gdansk (2006)	30 Jun 1998	30 Jun 1998	8	30.26	131.60	0.26
Zxxx/Krakow (2008)	15 Apr 1999	15 Apr 1999	14	20.75	90.12	0.67
Total for Poland			**101**	**170.87**	**743.27**	**0.60**
Total in Study			**2,504**			**0.51**

PATIENT FOLLOW-UP/RETENTION

A number of global clinical trials involving mortality endpoints require investigators to follow patient progress long after drug treatment has been discontinued. Experience has shown that it is important to work very closely with the sites and to provide them with the tools and the means necessary to ensure that patients are not lost to follow-up. In a recent severe head injury study, wherein the patient's status was collected and measured at 180 days, the study management team, through diligence with the sites, had only 2 out of 950 patients lost to follow-up at day 180. In a sepsis trial of over 2,300 patients, only 4 patients were lost to follow-up at 90 days post-randomization. It is important to provide staff with purpose and incentive when it comes to the collection of such data and to closely monitor the collection and timeliness of the follow-up patient visits in studies. Upon randomization, the study management team can provide the sites with a list of the projected patient visit dates for their reference in planning for all activities needed (see Table 12.3). This is a document that can be generated from most trial management systems and provided to the site as a fax or e-mail following (or with the confirmation of) randomization. The monitors should also have access to the same documents to use as worksheets for their site management responsibilities.

MEDICAL PRACTICES AND HEALTHCARE EQUIPMENT

Throughout North America, Western Europe, Central Europe (specifically, Poland, the Czech Republic, and Hungary), Israel, South Africa, Australia, and New Zealand, Western-style medical practices are broadly similar. This is especially so in the critical care and specialty units. Hospital ICU equipment (used to measure parameters needed for the inclusion/exclusion criteria) and preentry scoring, such as Glasgow Coma Score, SIRS (Systemic Inflammatory Response Syndrome) score, SOFA (Systemic Organ Failure Assessment) measures, and so on are fairly comparable. However, treadmill protocols and specialty equipment/measurements such as those used for ophthalmology evaluations, and so on, may vary greatly from country to country, making standardization essential. Investigators are aware of the need for standardization and information, and many institutions throughout these countries are interested in participating in global trials. In these situations, support from the institution and department heads is essential. During the prestudy evaluations, extensive questionnaires regarding the equipment and illustrations of the precise equipment expected and required are crucial. Outdated equipment may need to be replaced, making equipment expense a budget consideration prior to undertaking these trials. Where specific equipment will be used for either patient inclusion/exclusion criteria or for the study objectives, care should be taken to record and compare the type of equipment to be used, including manufacturer, model numbers, service maintenance, and SOPs employed at different institutions.

Although these institutions may have the requisite equipment, it is also important to understand the medical practice for scheduling the use of the equipment. MRIs, CT Scans, PET, and other high-investment equipment may not be staffed adequately twenty-four hours a day. If the clinical trial requires it, special arrangements need to

Table 12.3   Patient Tracking with Visit Projections

Patient Visit Tracking with Visit Projections
Study Name/Drug Name
Protocol Number: 123456
As of 30 October 2000
Country: UK

Investigator/ Site	Patient Number	Patient Drop?	Visit Number	Actual Date of Visit	Projected Visit	Out of Window
Bxx/Clayton (72)	0001	No	1	10 Aug 00		
			2	07 Sep 00	07 Sep 00	
			3	11 Oct 00	05 Oct 00	–6 days
			4		02 Nov 00	
			5		30 Nov 00	
			6		28 Dec 00	
			7		22 Mar 01	
			SF*			
	0002	No	1	11 Aug 00		
			2	08 Sep 00	08 Sep 00	
			3	12 Oct 00	06 Oct 00	–6 days
			4		03 Nov 00	
			5		01 Dec 00	
			6		29 Dec 00	
			7		23 Mar 00	
			SF*			
Vxx/Hull (75)	0001	Drop	1	01 Aug 00		
			2		29 Aug 00	
			3		26 Sep 00	
	0002	No	1	07 Aug 00		
			2	04 Sep 00	04 Sep 00	
			3	10 Oct 00	02 Oct 00	–8 days
			4		30 Oct 00	
			5		27 Nov 00	
			6		25 Dec 00	
			7		19 Mar 00	
			SF*			

*SF = Safety follow-up visit.

be provided for. In addition, whereas in some countries it may be standard care to follow up patients at three months with a CT Scan or other such measure, it may not be so for other centers or countries, due to limited availability of equipment.

As for procedures not requiring equipment per se, medical practices between countries still require consistent and standardized recording. Many rating scales not associated with equipment (such as the Hamilton Depression Scale [HAM-D], Rankin Scale, Karnofsky Scale, Hamilton Anxiety Scale [HAM-A], and various Quality of Life [QOL] scales) are often not commonly used outside of clinical research, thus many site staff may not be familiar with them or may not use them on a regular basis. It is important to ensure that the scales themselves are standardized for multilanguage and multiple cultures. It is important to ensure that these scales have been translated and *officially* standardized into the languages required before starting the trial. Additionally, training, standardization among the raters, and retraining and restandardization during the course of the trial are imperative to reach meaningful data at the end of the trial.

It is essential to realize also that the healthcare systems themselves differ in terms of private care, national health insurance coverage, public hospital management, and so on. A country's health policies and reimbursement schema often impact the investigator's ability to maintain control over the patient enrolled into a study. And, local health policies impact such measures as:

* The number of days in a hospital;
* When (that is, how long after the onset of the disease event) individuals are admitted to the special units for care;
* When and how "Do not resuscitate" (DNR) orders occur; and
* How intermediate-term care is handled (step-down units versus discharge to another facility or to home).

Local healthcare policies differ not only outside the United States but within the United States as well. In order to control some of these variables, it is advisable to increase the number of sites per country where possible, rather than adding more countries. Increasing the number of sites per country is more economical to manage and allows the data to be analyzed for spurious country-effects and comparability. Specialty units may serve more than one type of patient population and an understanding of the different types of patients entering the units during the selection process is advisable to ensure adequate space for the study population. As an example, ICU units in some institutions may serve general medical and surgical, as well as cardiac and neurological, patient populations. However, in larger cities and medical centers, medical and surgical ICUs often exist separately in addition to specialized cardiac and separate neurological critical care units. Some studies are designed to accommodate a variety of acute and chronic care areas of hospitals, such as emergency wards, surgical theaters, radiology units, ICUs, step-down units, and rehabilitation units or separate institutions. In such cases, it is important to assess the investigator's and his or her staff's ability to move freely between these units and to assess the hospital management awareness, support, and commitment to the requirements of the study. Additionally, it must be recognized that in different countries, a different type of investigator may be needed to identify and to conduct the trial. General practitioners are often the gatekeepers needed to identify patients who are de novo or in the early stages of various diseases, where in other countries, it may be more appropriate to go to specialists such as cardiologists, neurologists, and so on.

It also should be noted that the size of specialty units (generally measured in number of beds) is not always a good predictor of enrollment or compliance. Both large units and small units have enrolled patients and participated successfully in global trials. However large or small, they share key common features: a dedicated, committed investigator and a good team (however large or small) to work with.

INVESTIGATIONAL FEES

It is general practice in the United States and often the law in various European countries that investigational grants are paid only to the institution or to a special research fund rather than to the investigator himself or herself, since the investigators are generally salaried employees. In some countries, such as South Africa, the general medical society encourages standardization of pricing for various tests and services and when investigators are offered more than what is expected in their country, they can be offended as an ethical breach of their interest in participating. Investigators are sometimes looking for financial incentives outside of payments made to their institution, and therefore, it is important to proceed cautiously when negotiating grants and to know the compensation structure within a country before discussing grants. Obtaining compensation structure information may involve discussions with hospital management and the institution's legal departments.

Due to competition for the placement of research in hospitals and other study sites, pharmaceutical companies and/or CROs have, as an industry, driven the price of the investigator grants higher than ever before. National healthcare reimbursement schemes and institutions view investigational grants as necessities to help fund the healthcare systems. Clearly, demands on personnel time, especially with paperwork and "stat" requests from CRAs, driven by the sponsors' need-to-know on a timely basis, are costly. As an industry, we are ethically obliged to ensure adequate and appropriate compensation for demands as to safety concerns and compliance needs. Fees should be set to adequately and appropriately compensate investigators and staff at specific country cost levels plus a reasonable margin; arbitrary fees designed to pull investigators from competitors should be avoided. In all countries throughout the world, investigators who are interested in research and interested in the science of the study will participate at fair rates. When price demands are the issues prior to commencement, it is generally a sign to reconsider enlisting that particular site. Prices should be set around costs and costs should be very country-specific and calculated in the local currency (as opposed to one $U.S. amount for all). Rather than using currency exchange values and fixing the price amount, the value should be equitable (not necessarily equal). Toward this end, the principles of purchasing power parity (Chen 1995) should be considered when setting up financial agreements in global and multinational trials. The study management team SOPs should take into account the cultural, national, and local requirements. In a global or multinational trial, although it is wise to keep consistent practices and activities throughout, it is equally necessary to commit to managing the cultural differences from the start.

The following several examples illustrate differences that can occur between countries. In France, there are two types of contracts: one for the hospital and one for the investigator. The investigator contracts must also be submitted to the CNOM (Conseil National de l'Ordre des Médecins) in accordance with Article L.4113.6 (ex art. 365.1) of the Public Health Code of France. This authority evaluates whether the fees paid to investigators in France can be considered reasonable compensation, taking into account the investigator's workload. If the contracts for the investigators are all the same, only one submission needs to be made, which would include: the Investigator Agreement template in French, the list of French sites and investigators, the final protocol, the Case Report Form (CRF) or latest draft thereof, and the approval from the Comité Consultatif de Protection des Participants à la Recherche Biomédicale (or CCPPRB—the name of the IEC used in France). If there are different templates for the investigators or different budget amounts for fees, you are required to submit all of these documents as well. The CNOM answer should be received within one month (but it has been known to take longer). Experience has shown that these boards take their responsibility seriously and will indicate when the price for the study is ethically too high for the tasks required. This is a concern seen most often with naïve United States sponsors who require or request that all centers worldwide receive the same compensation in United States currency. In Italy, the sponsor must negotiate the contract with the investigator first and include a copy of the signed financial agreement with the ethics committee review material before the committee will review the study. In the UK, the National Health Service (NHS) is beginning to calculate costs for the procedures and overhead and the financial agreements are signed between the sponsor and the hospital, not with the investigator. In Spain, because the law regulates the financial responsibilities of the individual hospitals and institutes, each hospital has developed its own financial contract (in Spanish) that they use with pharmaceutical companies. When sponsors require control over contracts in Spain but the negotiation is the responsibility of a CRO, or when the contracts are written by the sponsor/CRO instead of the institution, it will often take at least two to three times longer to negotiate and resolve. Thus, if pharmaceutical companies and CROs are not sensitive to the national laws and requirements, they will be frustrated by time delays and mounting management costs.

CENTRAL RANDOMIZATION AND MEDICAL HOTLINES

Central randomization is costly and should be employed only when this method provides direct benefit over the more conventional methods of allocating therapy. There are three main situations when centrally controlled randomization is warranted: when the investigational product is limited in quantity or of concern, to ensure sophisticated stratification or randomization schedules are followed uniformly, and sometimes in randomized, open-label studies to decrease the potential for bias.

From experiences in many global trials, investigators have been very positive about their experiences with central randomization schemes, whether they use Interactive Voice Response Systems (IVRS) or more manual randomization setups. CRO vendors with experience in multinational or global trials should have experience with central randomization as well and should be cognizant of the best way to set up these systems and to interface central randomization into the drug supply management, data tracking, and project management tracking activities. CRO vendors with global experience often have a central randomization group within their organization and have worked with other vendors of these services.

Investigators and their staff can, in general, use such systems easily, provided the CRAs give appropriate training. In some countries, an actual test call is advisable during the initiation process. During the evaluation and site selection process, it is important, in some countries, to ensure that the hospital service will allow toll-free or free phone numbers to be called from within the unit. In other countries, investigators and CRAs may need to make special arrangements with the hospital director to ensure that such calls would be allowed. In the recent past in the Czech Republic, the national phone company would not allow toll-free numbers to be obtained from vendors outside the Czech Republic, such as the United States telephone company MCI and British Telecom (BT). In this case, the hospitals were provided with a line-item payment within the hospital grants for calls to the randomization center, hotline, and Serious Adverse Event (SAE) centers. In addition, the CRA provided local phone numbers (directed to the CRO national office) as a backup for the sites at all times. The CRO office would then make the call to the randomization center or hotline, if necessary.

In a 19-country, 280-site study recently conducted, investigators and their study staff managed the randomization and hotline discussions in English with few difficulties. The CRO monitored the process closely to ensure that calls were coming in from various countries and, at one point, it found that a number of investigators in Spain as well as their study staff were hesitant to use the medical hotline. Therefore, the CRO provided a special Spanish help line through the local staff in Madrid, who served as intermediaries, or where they were able, answered the questions directly.

Central randomization vendors should have the capability of providing IVRS in multiple languages and across time zones with adequate backup support to cover the hours required. This will often mean 24 hours a day, 7 days a week, and 365 days a year. It should be realized that some countries, such as Israel, conduct normal working hours from Sunday to Thursday and that global coverage will require multiple continents and time zones.

As for site preparedness, specific site personnel are generally provided with a small laminated copy of the script and the expectations of the call. These can be in the local language. It is advisable not to require responding with letters of the alphabet via the phone tone pad unless absolutely necessary, since many phones throughout the world do not have the alphabet on the phone keys and/or the alphabet is different for different languages.

Especially in global and multinational trials, experience has shown that a hotline service for information on all aspects of the study, from inclusion/exclusion decisions to dosing management to adverse event reporting, is a valuable and necessary aid. Information discussed during such calls is invaluable in analyzing site needs, understanding, retraining, and basic troubleshooting for these sites. In addition to the actual information (or message), it is valuable to analyze:

- The time of calls (a.m. versus p.m.);
- The date of calls (weekend versus weekday);
- The caller type (coordinator versus investigator);
- The number of calls per site; and
- The calls by country.

This information can be useful to assess enrollment, compliance, or administrative issues that may impact the success of the trial.

CENTRAL AND LOCAL LABORATORIES

Global clinical trials utilize data from either a central laboratory, the sites' local laboratories, or, in some cases, a combination of the two. It is the hallmark of clinical trial reports today to obtain good trend analysis data supported by good standardization of laboratory data. Such standardization is best reached generally by employing a central laboratory.

In severely ill patients, however, blood sampling can be extremely time dependent in the cascade of the illness events. In such cases, inclusion and exclusion criteria can only be based on local laboratory values, and the medical management of the patient can only be done ethically with such data. Labile samples such as hematology parameters and urinalysis are best managed locally. Some samples are drawn principally for good patient management and secondarily for the clinical trial. Central laboratory data are not available to the investigator for at least several days and should not be depended upon for clinical decisions.

Local laboratories in Europe and Central European countries, Israel, Australia, New Zealand, Latin America, and some countries in Southeast Asia, such as Thailand and Singapore, are, for the most part, up to the same standards as CLIA-certified (Clinical Laboratory Improvement Act) laboratories in the United States. To comply with ICH GCP, a local laboratory assessment SOP should require data from the laboratories accredited by a national or international regulatory agency. Or, if the laboratories in a country do not have access to such programs, during the prestudy selection visit, the CRA should investigate each hospital laboratory to determine their credentials (examining evidence of internal/external quality control programs) and the credentials of the laboratory director. If the CRA has doubts or requires further assistance, their quality assurance group can assist in the review process and/or visit the laboratory itself. For example, a few years ago, we investigated the possibility of conducting ICU studies in the Slovak Republic. After the prestudy visits and discussions, we advised against using sites in this country because the hospital laboratories, for the most part, did not meet the standards we would require for that study at that time. Alternatively, the management of central laboratory samples would have been difficult at best and more time-consuming from this country—an ineffective use of management time for the trial.

Transportation of blood samples from many countries to a central laboratory location generally requires more supervision and management than does transportation from sites in the United States. In some countries, such as the UK and France, the hospital laboratory generally manages the central laboratory sample handling, such as centrifugation, aliquot distribution, storage, packing, and ensuring collection by the appropriate courier. In other countries, CRAs will need to take a more active role to ensure that the investigator's staff has the appropriate equipment and storage capabilities, and that they can manage the packing and collection as required. In general, refrigerators and standard freezers are available to the study sites; however, when –70°C or other special requirements are needed, both the country and site conditions must be thoroughly investigated before initiation. As an example, last year we needed to arrange a local collection of samples that could be maintained at –20°C for up to two weeks, but then required –70°C until the batches were retrieved for the central laboratory. We identified one of the medical centers in the country to store all the samples for the other sites. After arranging the initial logistics, compliance and cooperation from all the sites were excellent. These actions were undertaken prior to study initiation and our local office recognized the potential problem early in order to provide alternative solutions.

There are many central laboratory vendors who have participated in global trials, and care should be taken in evaluating their experience with each country expected to participate in the trial. It is especially important to discuss the courier service(s) that are employed in each country and their ability to transport samples over weekends and holidays, since country holidays vary, creating important timing issues. Courier services management should also discuss how refrigerated samples are monitored and the plans to prevent delays in transport, especially in extremely hot climates or during times of temperature extremes.

In almost every hospital study outside the United States and Canada, when dry ice management is required, CRAs in the various countries have had to manage the process in the past. Today, some of the central laboratories are handling this activity by providing a dry ice shipment prior to collection of the samples (generally the next day). This type of logistics is a challenge across countries and CRAs should maintain close contact with the sites during the process or schedule their time to be present to assist the sites when appropriate.

PHARMACIES

In general, each country handles a site's drug supply management differently. A good number of hospitals and medical centers in the European Union utilize their hospital pharmacy for the storage, distribution, and sometimes randomization of the trial supplies. The UK, France, Sweden, and the Netherlands are all examples of such countries in this category. Routinely, the pharmacy is visited separately during the prestudy and initiation visits. A number of these institutions have a clinical trial pharmacist and a separate area for the storage of supplies and maintenance of trial records. However, this is not the case in many hospitals in Germany or Spain, for example. In these countries, the investigators keep the trial supplies in their department or in their office areas. It may be necessary to supply them with refrigerators and to diligently monitor temperature recorders. Locked storage facilities for ambient supplies are generally not a problem; however, access to these areas for continued resupply to the study patient and ensuring return accountability and record keeping can be problematic. There must be close monitoring by the CRA. Otherwise, it is very important, during the site selection process, to discuss the possibility of recruiting a research, or other carefully selected, pharmacist onto the investigator's team.

IMPORTING INVESTIGATIONAL PRODUCTS

Each country maintains specific laws and regulations on the importation of clinical investigational products. These laws must be strictly adhered to and planned for prior to the start of the study. The procedure for obtaining an import license can range from automatic approval upon receipt of the Ministry of Health approval to begin the trial, to the requirement for special forms and licenses for all drug supplies throughout the course of the trial or a special permit for each shipment to each site throughout the entire study.

Drug shipment planning must take into account the need to transport temperature-sensitive materials in a very timely and controlled manner to numerous countries within and outside Europe (including shipments to Israel, South Africa, Australia, Central Europe, South America, Southeast Asia, and so on). When appropriate with a new project, test shipments can be sent to validate the procedure before the supplies are available.

It is important to note that one should not expect investigators and/or pharmacists in various countries such as Poland, South Africa, and so on to be listed as the importer of clinical trial products. This is because the investigator or pharmacist may need to be available to the customs officers—sometimes physically—in order for the investigational products to clear customs. Although courier companies claim that they will handle the paperwork, it may be necessary to obtain a specific "power of attorney" from the sites for the couriers or other agents to clear customs. Thus, it may be more productive to seek a global or multinational CRO who can manage the importation of the product into the country. CRO offices are generally located in the capital cities where customs offices are located and they generally have the capacity to temporarily store the products in their offices (sometimes only overnight) and to arrange for appropriate transportation to the site. In France, only a pharmacist may import drug supplies for clinical trials and, for each shipment, a new (separate) import license is required.

A CASE STUDY

In November 1997, a pharmaceutical company had been conducting a double-blind, actively controlled, clinical trial for more than one year in patients who developed neutropenia due to systemic fungal infections. The company had less than one fourth of the expected enrollment for the program. The patient population consisted of hospitalized, severely ill patients with underlying disease that caused them to be immunologically suppressed or who received medication that caused suppression of their white cell defenses. Due to slow start-up (because of regulatory issues and questions, and a lack of general enrollment in the fifteen-country, Pan-European study), the pharmaceutical company went to a second CRO to solicit advice and assistance to complete the study in a timely manner. This study could only be conducted in hospitals, generally in their intensive care unit. Thus, there were a limited number of possible locations in any one country. The first solution was to critically review the inclusion/exclusion criteria and the definition of the study patient population. This review did not identify any possible changes because of regulatory considerations and standards of care. After an extensive feasibility analysis in countries not yet involved, regulatory/importing issues were reviewed and these activities resulted in adding eighteen sites in three new countries on three continents (see Figure 12.3). The original goal of adding seventy-five additional patients was reached six months after the first

Figure 12.3   Case study flowchart: Expanding a Pan-European study into a global design

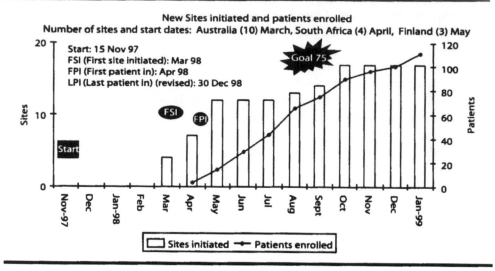

new site was initiated, and additional patients enrolled contributed to the total number of patients required by the protocol. The primary challenge for this study was communication, which required coordination with the pharmaceutical company, the other CRO, contract manufacturing and shipping organizations, the data management CRO, and internally with sites and CRAs on three continents. Successful completion occurred only as a result of the diligence of and overall good project management by the project manager and her ability to build relationships with these groups.

REFERENCES

Chen, B. 1995. Long-run purchasing power parity: Evidence from some European Monetary System countries. *Applied Economics* 27(4): 377.

Code de la Santé Publique. 1994. Law no. 94-43. Relative à la santé publique et à la protection sociale. Art. L 4113.6 (ex art. L.365.1). *Journal Officiel de la République Française* (France) 15 (19 January): 960.

Nigel J. Dent

Outsourcing and the Use of Contract Quality Assurance

We live in an ever-changing world and, as we move toward the next millennium, the plethora of large and small multinational companies producing pharmaceutical, veterinary, chemical, and other like compounds will probably reduce in both size and number.

Over the latter part of the Twentieth Century, we have seen a growing reluctance by companies to employ large numbers of staff, an increasing number of company mergers and acquisitions, and, following those activities, a downsizing of staff with a commensurate increase in outsourcing resources.

The virtual company is a logical extension of these trends. Here, a small core of key people with the ideas, and naturally the finances, puts together a clinical development program that is totally outsourced. All key members of the clinical development and preclinical development team reside outside the company, and their services are procured on an as-needed basis. The financial implications of this company structure are overwhelmingly positive. Gone are the large numbers of permanent staff requiring salaries, fringe benefits, company pensions, and insurance benefits; the expertise is hired as required and then dispensed with. The other benefit is that the company can obtain the services of very high-quality people with vast experience in the industry and, most importantly, respected names, who have worked for many multinational companies and have now set themselves up as independent contractors.

In addition to having a wealth of past experience, these people now work for several multinational and varied companies and therefore have a breadth of experience regarding the standard activity within the industry. Thus, while maintaining total confidentiality, these professionals can assist the company that currently retains their services to revise or review its situation, standards, working methods, and practices to ensure the company's compliance with the industry norms set by the lead company in a particular field.

This chapter examines the use of external resources to achieve internal quality, and covers the consultant Quality Assurance (QA) person, auditor, or, for that matter, quality controller for Good Laboratory Practice (GLP), Good Clinical Practice (GCP), and Good Manufacturing Practice (GMP).

CONSULTANCY SERVICES

Consultancy services offer a range of services, depending on the clients' needs. In its broadest sense, a consultancy service offers a expert advice where a particular complex problem is to be solved. However, maybe the client's requirement is to have an extra pair of hands and additional resources at times when its existing QA unit is overloaded and, therefore, the client really needs a well-trained auditor rather than the breadth and depth of the consultancy's experience as a whole.

Over the past few years in the UK, and to a lesser extent in Europe, the outsourcing trend, led by the United States pharmaceutical and allied industries, has rapidly engulfed the pharmaceutical and bioscience industries in these countries. According to the old cliché, if you don't have the expertise, then become a consultant. Today, the cliché is not only old, it is untrue.

All too often, consultants are seen as people who picked up a set of guidelines or regulations, read them, and then purport to be an expert in that particular field. Any particular consultant who follows this method can rest assured that, within a very short time, their workload and reputation will clearly indicate a lack of knowledge. In every instance that the author has come across, the client has an exact knowledge of what is required and the time that it should take to achieve this aim. If this differs from the services offered by the consultant, he or she can expect a very swift end to that contract and any prospect of repeat business. Contrary to public opinion, the consultant faces tremendous pressure to deliver the goods in the shortest possible time and with the utmost efficiency. To these ends, the consultant must know his or her subject in finite detail and should not stray from the particular discipline he or she has developed over many years prior to becoming an independent.

A consultant can bring in new ideas, new approaches, and a wealth of experience gained across several companies, countries, and disciplines. Now, consultants in the many branches of pharmaceutical and other similar disciplines offer a ready "library and catalog" of information that can be bought and equally rapidly dispensed with when the job has been satisfactorily completed.

Good Laboratory Practice (GLP)

In the areas of GLP, the consultant can offer several advantages. Where a company is new to the area of GLP, the consultant, using his or her previous knowledge of the local inspectors and their techniques, can put the company on the right track by interpreting the regulations or guidelines according to the local inspectorate.

Frequently, a company has a basic knowledge in a particular area and has worked for a few months to achieve the principles of GLP but lacks the time, money, or resources to employ a full-time QA person or set up a QA unit (QAU). Here, QA can be contracted in or a person identified within the organization but sufficiently divorced from the day-to-day experimental procedures can be trained by a consultant to act as a part-time inspector in conjunction with regular visits by the consultant.

Some companies are employing consultants in a preregulatory inspection-visit guise. Frequently, a company has a QAU and thus believes it operates in full compliance with the principles of GLP. The company aims to apply for, and to achieve, compliance through a local regulatory authority inspection and then truly to claim compliance and have the all-important documentation to prove that they have had a satisfactory inspection. In some European countries, and particularly within the UK, the GLP Monitoring Authority of the Medicines Control Agency (GLP MA) offers an advisory visit. However, many companies feel it is more beneficial to obtain an advisory visit by somebody who will be completely independent and will not subsequently inspect the company's facilities with prior knowledge of its noncompliance points. To this end, the consultant can spend from one to five days with that company, going through all aspects of documentation, experimental procedures, auditing, QAU operation, and archiving procedures as if the consultant were a regulatory inspector.

It should be stressed at this point that consultants cannot set themselves up in any way as a quasi-independent regulatory authority. In many instances, the consultants themselves have been through several inspections by different countries' regulatory bodies and therefore know the format and the questions along with the laboratory areas that are likely to be involved during a regulatory inspection. The standard technique used by many consultants is the Organisation for Economic Co-operation and Development (OECD) checklist, published in the official journal, and also Document 88/320 of the European Union (EU), published in the journal of the European Communities. These documents clearly indicate what an inspector would look for when conducting an inspection under the auspices of the OECD principles of GLP.

Whatever route the company decides to have the consultant use, it will be looking to get three things in return: first, a summary of the company's status with regard to GLP; second, a detailed report outlining areas of noncompliance and recommendations for implementing procedures to achieve satisfactory compliance; and third, a timetable and a schedule of regular meetings with the consultant to achieve the aims and objectives outlined in the initial report.

Good Clinical Practice (GCP)

The ways in which a consultant can be helpful in the GLP and GCP areas are similar. Again, the consultant would be most beneficial in providing on-site audits for multisite, multinational clinical trials. In many cases, hiring a consultant for those days required, dispatching them to the country or center concerned, and then using their expertise for all site audits for that particular compound in that particular series of studies is much more cost-effective. This ensures continuity of approach and inspection technique, while allowing the company QA person or persons, if the company has any, to remain on-site and achieve the other aspects of QA auditing. One other area in which consultants can be particularly beneficial within GCP is in the audit of the company site final report. Priorities often change very rapidly and it is not unknown for major clinical trials and major reports of completed clinical trials to drop suddenly onto the QA manager's desk, requiring many problems to be dealt with, all in the three days running up to Christmas! A consultant can be brought in for that specific piece of work. It is even more beneficial if that consultant has previously worked with the company in areas of site and investigator auditing because he or she will be familiar with the format of the company's QAU, the type of study and data, and, in some instances, the investigator who has produced the data for audit contained in the final report.

Two other aspects of contract consultancy in GCP areas are the preplacement visit (to ascertain the status of *laboratories*, both for clinical and GLP, prior to conducting a clinical trial) and the production of company Standard Operating Procedures (SOPs). In addition, contract research organization (CRO) auditing is of particular importance if a Phase I study is involved.

The task of assessing laboratories and compliance can include evaluations of the sponsor pharmacy; the clinical trials' dispensary; the investigator's clinical pathology, cardiology, ophthalmoscopy, and phlebotomy laboratories; and specialist laboratories attached to the clinic, which may include X-ray, CAT scans, and so on. In addition, inspections could be made of the analytical drug substance measurement laboratories, the pharmacokinetics laboratories dealing with the analysis of the analytical data, and many other ancillary laboratories involved in any clinical trial.

The consultant is often required to assess Phase I facilities for their ability to conduct particular clinical trials and, frequently, the consultant is dispatched to the appropriate Phase I unit to give an overall indication as to how that unit functions and whether or not it could conduct the clinical trial on behalf of the consultant's client.

Consultants are often very beneficial in helping to generate SOPs or to oversee them. Consultants have access to many companies' documentation and, although under a strict confidentiality agreement, they are often in the very fortunate position of having seen both good and bad documentation, enabling them to advise a company on compliance with the industry norm.

The consultant's activity in assisting with SOPs is one of several documentation areas in which a consultant can benefit the client. Because consultants are privy to many types of documentation, it is natural that when they see good ideas, they incorporate them into their scrapbook and use them to help new companies set up facilities and produce SOPs and other documentation. One could argue that the use of consultants is one way of enabling many companies to take a standard approach to documentation, thus aiding the inspectorate, since they will see the same types of information, documentation, and SOPs provided from company to company and country to country and this, in itself, will help to standardize the clinical trial situation.

Good Manufacturing Practice (GMP)

The benefits that a consultant can bring in GMP, in many instances and in the broadest terms, are similar to those a consultant brings in the GCP and GLP areas. Naturally, in the production environment, there is a much more hands-on approach and, in these areas, the consultant must have a very thorough knowledge of the regulatory requirements and the inspectorate and their attitudes toward production plants. But the consultant must also be aware that his or her recommendations could involve the company in enormous sums of money setting up or changing a plant and, therefore, the consultant must be 150 percent sure about what he or she is advising the client to do.

Computers

While the good practices have been addressed, the use of computers within the various scientific disciplines needs to be further addressed. All those good practices (GLP, GCP, GMP) interact and, often, computers perform and control the same function across these good practices (spreadsheets, data handling and management, control of equipment, etc.). Computers in these fields must demonstrate validated systems and that they are "controlled." The old adage "garbage in, garbage out" is as pertinent here as in any area where computers are used.

Validation, compliance with particular guidelines or regulations (electronic signatures, software or hardware validation), requires a specialist. It is in this area where computing/computer consultants can be of major assistance. A consultant who is aware of the requirements, the marketplace, and problem-solving techniques and is working with the company IT department can rapidly solve major company problems.

CONSULTANTS AND THEIR USE BY COMPANIES

One benefit of using consultants is that they are expendable commodities. In many instances, companies require areas of expertise that consultants can fulfill and company staff can be trained to produce. However, it is not cost beneficial to take a person on board and train him or her for a specific short-term topic or task. Where this occurs, consultants with guaranteed expertise in the areas required can be bought in, or brought in, used to complete the task, and then dispensed with.

Consultants do not require private pension schemes, Christmas bonuses, bonus packages, holidays, or sick leave and they are always there when required by the company.

These days, QAU staffing in the scientific fields is kept to the most economic minimum of staff and senior management—critical mass is the current phraseology. Frequently, however, there are peaks in the workload that put the staff under extreme pressure and quite often they are unable to cope with the demands placed on them by management. In these particular times, consultants can be brought in to alleviate the staffing problem. If particular consultants have been used before, they can be slotted immediately into the area that requires extra staffing. However, even if the consultant has not been used by the company before, he or she is unlike new staff members and unlike workers brought in on a temporary basis from an agency. Here, it may be at least six months with junior staff and at least three months with more senior staff before they are actively producing guaranteed quality work according to the company standards. Unlike a new "trainee" who will take time to settle in and understand the company's workings, consultants have a considerable range of expertise and many years of experience and will be able to operate from day one, hour one. In addition to the other aspects outlined, the trainee with a few years' experience moving into a new job from another company is not as prepared as a consultant whose daily task is moving from company to company and country to country, fulfilling just that need. The consultant has been absent from the security of full-time employment for several years and therefore can readily adapt to the changing environments between companies.

Having examined the various roles the QA consultant can fill within companies, this chapter will now move on to a consideration of the pitfalls and benefits of using consultants. These are outlined for both the company concerned and the consultant.

CHOOSING A CONSULTANT

Qualifications

There are many ways that consultants can obtain qualifications. Naturally, consultants should possess academic qualifications and specialty qualifications in the discipline in which they offer services. These qualifications should be checked back to a CV. In these days of assurance, it is always good practice to go back to the original document to ensure that any stated qualifications are actually current and valid.

From a scientific viewpoint, the qualifications should be carefully checked to confirm the length of time the consultant has been in the discipline and to verify that those areas of stated expertise are clearly backed up by experience in the appropriate company with the appropriate tasks and responsibilities. A thorough check of the consultant's CV and references should be made.

It is very dangerous to rely merely on the consultant's company brochure and CV as a guarantee that a consultant is able to fulfill your requirements. The author suggests that, for any contract, short- or long-term, a personal interview be undertaken and, if necessary, a clear examination of the consultant's qualifications, along with a face-to-face review of the activities required by the client.

Interviewing Consultants

Naturally, the best consultant is recommended from another company or a group who has used that person. It is often difficult for consultants themselves to recommend colleagues because, although a colleague may be well-known in the industry, it is a difficult task to recommend somebody for a job if you have not worked with that person directly or know their practical and actual capabilities. While a person may be well-known and well-respected, it is not until a task is undertaken and a satisfactory conclusion arrived at that any particular task or activity can be described as satisfactory, meeting the client's requirements.

Certainly, a live interview is a critical part of the process of choosing a consultant. Prior to the interview, you should fully read the consultant's CV, review his or her record of training activities and other documentation, and check the appropriate references, either from other companies who have used the consultant in your particular area of need or those who can provide a general overview of how the last contract was completed by the consultant. Once the background documentation has been reviewed, the consultant should visit the premises and participate in a full and frank discussion regarding the task at issue. Questions about how the consultant would solve the relevant problems should be posed. The interview should be as thorough as it would be if the person were applying for a permanent post. Following a satisfactory interview, the consultant should be asked to produce a very accurate time plan, action plan, and estimate of cost and, where applicable, to either supply extracts of or make available his or her SOPs for conducting that piece of work. All consultants in any discipline should have comprehensive and regularly updated SOPs.

Because certain good practice disciplines are emerging, there is now the possibility for consultants, along with all QA personnel, to obtain academic qualifications in the good practices and this is an area that prospective clients might consider. However, a diploma in QA is no guarantee that a consultant will be able to fulfill your job requirements and thus, a diploma alone should not preclude the detailed selection and review process described previously. Even with academic qualifications, however, the practical nature of the consultant's past work should be fully reviewed since a consultant's hands-on ability, rather than his or her paper qualifications, is what a consultant will need to get a particular job done.

Insurance and Contract

The last step in choosing a consultant is reviewing his or her insurance and contractual requirements. All consultants will be covered by some level and type of insurance and, again, this should be reviewed in light of the job that will be undertaken. Professional indemnity insurance, travel insurance, and health insurance would be the three main types of insurance that should be reviewed by the prospective client.

Many consultants, while wishing to have as much insurance as possible, find it prohibitive to purchase some kinds of insurance because the overall premium charged for a particular coverage prevents the insurance from being a viable proposition. For a one-time consultancy activity, insurance premiums may well run into thousands of dollars but only provide a level of coverage that, in the event of a claim, would not be sufficient. The client must therefore consider what insurances it is looking for and the reason it is looking for the insurance. These considerations should be discussed with a prospective consultant.

All consultants would have their own company quotation and contract and, again, this should be reviewed by the potential "employer's" legal department and modified to the satisfaction of both the consultant and the potential client.

PITFALLS AND BENEFITS OF USING CONSULTANTS

As in all comparative situations, it is beneficial to deal with the pitfalls first, so as to paint the bleakest picture, and then go on to the benefits. At least from that point of view, the reader will be left on a high note indicating, from the consultancy point of view, why you, as prospective clients, should employ consultants.

Pitfalls

At minimum, there are six tangible negative aspects of hiring a consultant working within industry.

1. **Cost.** Obviously, to employ a consultant is, in the short term, considerably more expensive than employing a staff member for the same period of time and the same proportion of work. The client must realize, however, that this is a onetime expense, that a consultant is brought in for a particular reason (staff shortage or a need for immediate expertise), and therefore the need is great and the remuneration must be paid accordingly.

2. **Availability.** As with all contractual arrangements, the consultant is usually required at a moment's notice. From the point of view of the consultant and the company, this is a pitfall.

   These days, management rarely thinks in advance of the requirement for instant expertise; usually somebody calls for a consultant when the most dire consequences are about to happen. Here, in the author's opinion, many companies seem to think that consultants sit in their offices with their head in their hands waiting for the phone to ring or the fax or e-mail to start chattering requests for them to fly immediately to Oslo to fulfill a site audit consultancy need. This is a distinct pitfall for the consultant, and if his or her inability to attend to this task is a ruling factor, then this obviously is a pitfall to the company. More advanced planning and liaison within companies would alleviate this pitfall.

3. **Language.** This can be seen most clearly as a pitfall for the consultant. When called upon to visit a foreign country to carry out work, a consultant is confronted immediately with a pile of handwritten raw data to be compared with a pile of handwritten or typed reports, both in the local language. While the concept of auditing is usually comparing like with like, and numbers are the same in any language, language does often pose an immediate complication in ensuring that an accurate and independent audit is carried out. There are obviously ways that these problems can be overcome, but the need to have someone translate and point out where data can be found sometimes prevents a truly independent and unbiased audit, in the author's opinion. The problem is extremely complex and difficult to overcome and, short of having a multilingual consultant aware of all the scientific terminology in the local language, this is something that most companies are still wrestling with.

4. **Commitment.** Both the consultant and the company can perceive that the other party's commitment is less than it would be if a permanent member of staff were concerned. The consultant is brought in for a period of time and, in that short term, he or she knows that the job is, more than likely, one that nobody else wanted to do. Perhaps it is a very mundane task and the consultant is aware that, after this task is completed, he or she will be off to another assignment. Perhaps a client believes that, given the circumstances, a consultant is less committed than somebody who is following a particular experimental study design from start to finish.

5. **Confidentiality.** This may be seen as a pitfall from the point of view of the company. The consultant will move from company to company and, naturally, will have access to very highly confidential documents. Some may result in a patent application; some may address the development of a new drug substance or similar matters. To alleviate this concern, all consultants and consultancy companies should sign a confidentiality and secrecy agreement immediately and work along the same conceptual lines as any CRO. Here, total confidentiality is given to each party and to every product and with every consultancy. Obviously, with the varied tasks set before the consultant, he or she will be party to many different aspects of confidential information. This must be discussed at the onset of the consultancy and, if the client feels that there is a threat to its particular environment, then this is one of the overall points for consideration in hiring that person. If confidentiality is seen as posing a problem in the overall project, the client should proceed only if it is happy with the various techniques and agreements drawn up.

6. **Acceptance.** It is quite often difficult for the staff and the regulators to accept a consultant. The intrusion of this so-called expert, brought in for a limited period of time, may be seen as countermanding the productivity of the scientific team. Again, as in all aspects of QA, the acceptability of the technique, the person, and the operation can only be dealt with using people skills, which the consultant should have already adopted. Regarding regulators, within Europe and, for a long time in the United States, inspectors themselves have taken consultants on board to help in dealing with specific problem areas where the regulators have little or no expertise. As a result of this trend, the inspection team should now accept consultants as at least a necessary evil. Through the agreement of mutual acceptance of data, reciprocity, and exchange of information within OECD countries, these trends can only lead to the general acceptance of consultants by regulatory personnel.

Having painted the bleakest picture possible when employing a consultant, perhaps a review of the benefits of employing such a person is in order.

Benefits

It is not surprising that the benefits vastly outweigh the pitfalls of employing consultants. Here, nine topics that detail the benefits to a company are presented. Obviously, any employment of consultants benefits the consultants as well.

1. **Expertise.** Undoubtedly the most beneficial aspect of employing somebody in a consultancy capacity is that his or her expertise can be readily determined prior to engagement. A consultant's main selling point is his or her length of time in the particular discipline or within the particular purported area of expertise. Usually, consultants have been in the industry and involved in their particular discipline for many years; their professional image and qualifications have gone before them to act as a ready publicity agent. A company can rarely obtain readily the exact area of expertise needed by obtaining new staff members or moving people from discipline to discipline. Employing a consultant,

where it is well-known that that person will be able to fulfill that function immediately on entering the company, is thus a very positive approach.

2. **Availability.** A consultant can be brought in or, as already stated, *bought* in on an as-needed basis. It is sometimes preferable to design a length of time within the experimental period when it is known that staff shortage will become a critical factor and a consultant can therefore be employed to cover that period of work. Many time-scheduling computerized software packages are available and this is one area that can readily be dealt with. Again, because this person is being utilized for a specific task, when that specific task has been completed, the consultant can be dispensed with and is not an overhead.

3. **Overheads.** A consultant can usually be funded from a general overheads budget rather than from the salary and wages budget, which is very often closely monitored and controlled by senior management or a financial department. Moreover, a consultant does not require the provision of private health care, National Health or similar health protection insurances, allocation from funds for pensions, or insurances to cover sickness and ill health— these are all provided by the consultant for the consultant. On these matters, a consultant is never sick, does not have holidays, and works for the whole period of time that they are in the company's employment. For these reasons, this type of employment is very cost beneficial. Overheads, particularly those seen from a staffing point of view (often classed as unseen overheads), are not an issue with a consultant. When a consultant has quoted a job, this is the final figure paid, allowing all budgets to be met.

4. **Tasks.** In general terms, the consultant is an all-purpose workhorse, willing to take on any task. In many instances, consultants are given those tasks considered mundane, boring, long-winded, and/or complex. It is, perhaps, unfair to say that a consultant is willing to take on any task; each task carries a price and, in many instances, the price fluctuates to reflect the consultant's desire to take on the job. In most instances, however, the consultant views taking on these tasks as a challenge, whereas the facility's staff would be very loath to take on "yet another carcinogenicity report audit." The consultant, on the other hand, is quite happy to perform the audit since, in working for the next client, he or she may well be auditing a clinical study or something totally different. Consultants can take on the more challenging audit and vary their daily tasks accordingly. The other overruling aspect is that, because at the end of the day they submit an invoice and are paid for the task, to a certain extent a consultant's willingness to take on any task expands with the financial reimbursement.

5. **Cost.** The cost benefits of using consultants have been well-covered in this chapter. It is definitely less expensive to pay a consultant a set fee for a limited project than to pay the salary and benefits of a full-time employee. The cost of a consultant is a negotiated charge that fits in with budgets and, provided closed contracts are usually undertaken, no hidden surprises remain for the client.

6. **Contacts.** In addition to the expertise that the consultant brings, one of the other fringe benefits is the contacts a consultant will have, or is able to make. Consultants work for more than one employer and, therefore, there are opportunities to make many contacts in various disciplines. Quite often, if a specific task cannot be fulfilled by the consultant, a friend, or a friend of a friend, is able to satisfy the client's additional work requirement. The benefit of this is that it becomes a recommendation by somebody you have employed on a professional basis and, to maintain their reputation, they are only likely to recommend somebody who is of a similar professional standing and is well-known to them. However, a thorough knowledge of their ability must be a key point in any recommendation with preferable firsthand practical experience in working with them. If this is not the case, then a reliable recommendation becomes a difficult task. Today, you will find that a consultant is often an associate and, therefore, pools of expertise can be found where a complete project can be undertaken by these external experts.

7. **Commitment.** You may remember that commitment was one of the pitfalls dealt with earlier in the chapter. It was described there as a possible adverse effect because of the possibility of the consultant having a lack of commitment to the overall project. However, it could be argued equally well that one of the benefits of having a consultant is that he or she is under contract, with an accepted quotation, guaranteeing the work pattern will be carried out to the fullest and best ability of the consultant. This ensures that either repeat work is possible, hopefully guaranteed, or that their professional standing can be further enhanced by recommendations from the company currently employing his or her services to another company who finds itself in a similar situation.

8. **Reporting.** In general terms, the prime problem with any aspect of quality assurance is the consultant's ability to handle the workload—the ability to turn reports around and to ensure that the action points are adequately dealt with to enable the report to go out within the critical time frame or window allocated by project planning. In a consultancy situation, this will have been dealt with in the overall appraisal of the task and sufficient time allowed for reporting to take place within that contractual time period. When one job is finished, additional consultancy will be required by another company and another task. It is therefore imperative, and in the consultant's best interest, to ensure that the report is prepared and dispatched within at least seven working days from the completion of the task if this is an on-site or general inspection. Obviously, if it is a report audit, then sufficient time will have to have been allocated and built in to ensure that the report is left with the client for discussion with the relevant study personnel prior to the termination of the contract.

9. **Independent Voice.** Very often, a consultant's presence for a particular topic has been carefully thought out by management who has a particular party line that they wish to be given by a person not representing the company. The consultant, by his or her presence, training role, or inspection technique, can reassert the company line but, because he or she is not a company

employee, this perhaps lends a little more credibility to the purported reasons that a particular action is considered necessary and thus will fulfill the general requirement of the company in a particular aspect of compliance that the company has found so difficult to implement earlier.

To summarize, this chapter has detailed, in a noncommercial sense, why it is beneficial to obtain the services of a consultant in a QA role in one of the aspects of the GXPs. This chapter has reviewed in general terms what the consultant can do for a company and why a company should turn to that person, not just as a last resort, but as an extension of its own professional image and the capabilities of its own research team. Attempts have been made to balance out the role of the consultant, and the desirability for a client or company to employ a consultant and the pitfalls and benefits from the company's and from the consultant's viewpoint have been outlined.

This balanced picture, therefore, should suggest the overall benefit of employing such a person to work in the pharmaceutical, bioscience, and allied industries to help out in the overall total commitment to quality that all companies now wish for.

REFERENCE

OECD. 1998. *OECD Principles of Good Laboratory Practice (as revised in 1997).* Series on Principles of Good Laboratory Practice and Compliance Monitoring, No. 1. Paris: OECD.

Gillian Gregory and John Glasby

Contracting Out Regulatory Projects to a Regulatory CRO

Regulatory affairs are no exception to the trend to increase the amount of work outsourced to Contract Research Organizations (CROs). By grouping together CROs, however, it is easy to fall into the trap of thinking that contract research organizations are all of one type, just as it is easy to think that Europe is one country or that regulatory agencies are all of one mind. Unfortunately, none of these statements is true.

In the clinical field, the rules and standards applied are broadly similar for different products in different countries; indeed, if the standards applied were to differ, it would create problems in gaining acceptance of the data by the client or, ultimately, by the regulatory agencies involved. The clinical CRO's objective is therefore to apply the rules of Good Clinical Practice (GCP) and Standard Operating Procedures (SOPs) as rigidly as possible. Thus, all projects will tend to follow a pattern allowing a skilled and experienced CRO to achieve the various milestones quickly and efficiently.

As a field, registration is not nearly so uniform; there are different styles of regulatory agencies, taking different approaches to the application's assessment, such that each project taken on will be different, requiring the application of a different set of criteria. Even in terms of the final dossier prepared for submission to the authorities, a single worldwide format or content is not yet possible. At present, there are developments with respect to introducing a common technical document, applicable to the United States, Europe, and Japan, which should be in place in 2003.

CHOOSING A REGULATORY CRO

Regulatory Systems

Before considering the regulatory service a CRO can provide, it is necessary to briefly discuss the different agency approaches, since these play a vital part in the selection procedure.

In general terms, there are three regulatory agency approaches in the world: the United States Food and Drug Administration (FDA) approach, the Japanese approach, and the European approach.

The FDA is a large organization with professional reviewers who approve or deny a variety of applications, from those seeking permission to conduct clinical trials to New Drug Applications (NDAs). The regulatory process is a continuous one, from the Investigational New Drug (IND) application to that of the NDA, with considerable discussion and negotiation between the company and the FDA during the process. The United States requires FDA approval of clinical and other key protocols (such as those for the formal stability test before the studies take place). The process is expensive and time-consuming, but if it is carried out properly, it should steadily progress to approval without unpleasant surprises.

Japan evaluates new drug applications on the basis of quality, safety, and efficacy. However, the Japanese file is unusual because much of the data in the application is published prior to submission. Japan has recently changed its assessment procedure. Until November 1999, there were three regulatory review steps after the initial Evaluation Center's review (i.e., subcommittee, special committee, and executive committee in the Central Pharmaceutical Affairs Council [CPAC]). In November 1999, CPAC was reorganized and the subcommittees abolished. The reorganization was intended to shorten the regulatory review time from eighteen to twelve months after April 1, 2000. In order to meet these timelines, the Evaluation Center may advise withdrawal of applications more frequently than before. To avoid this, companies are strongly advised to use the Kikoh Consultation System during the development phase. Although consultation is not obligatory, in practical terms it is essential. Companies consult with Kikoh on clinical trial protocols, development plans, and other issues. There is a charge for each consultation.

The International Conference on Harmonization (ICH) guidelines have been particularly useful in Japan, helping to create an openness that is allowing Japan to become much more involved in international pharmaceutical development. As in all countries, but particularly so in Japan, it is important to have appropriate local contacts who can act as resident experts, providing the country-specific input to be considered in global development plans. As local experts, Japanese CROs can play a vital part in new drug development for the foreign developer.

The European approach, whether through one of the European Union (EU) procedures or through local application, is less interactive than in the United States and thus, a little less precise and certain. Under the European approach, there is limited but increasing discussion of the development plan or protocols with regulatory authorities before the studies commence, since there is, at present, no uniform clinical trials approval process. A European Clinical Trials Directive was introduced on May 4,

2001, but this is unlikely to significantly move the European system toward the much more continuous system used in the United States.

European Centralized Procedure

In Europe, EEC Council Regulation No. 2309/93 created a centralized community procedure for approving medicinal products, shown in Figure 14.1. The procedure requires the submission of a single application, provides a single evaluation, and grants a single authorization, giving direct access to the single market of the Community. Biotechnology products, listed in Part A of the Regulation, must follow the centralized process; the procedure is optional for the innovatory products in List B. The process works on the basis of a 210-day review period, although with clock stops to respond to questions, it is realistic to think in terms of twelve to eighteen months for completion. The applicant can express his or her preference for a rapporteur and corapporteur, but the final decision rests with the European Medicines Evaluation Agency (EMEA). The procedure requires the provision of a single trademark throughout the EU; because the European Commission does not wish to facilitate the partitioning of the market, any requests for multiple applications must be justified to and approved by the Commission in advance. The procedure is working well, but there is a growing concern about the number of withdrawals, the majority of which are on the basis of efficacy and safety.

European Decentralized Procedure

The other major European regulatory procedure is the decentralized procedure or mutual recognition procedure, shown in Figure 14.2. From January 1, 1995, a pharmaceutical company wishing to market a drug in more than one Member State could

Figure 14.1   Centralized procedure for EU product approval

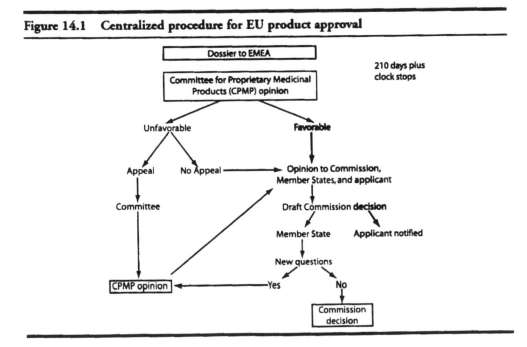

Figure 14.2   Mutual recognition procedure

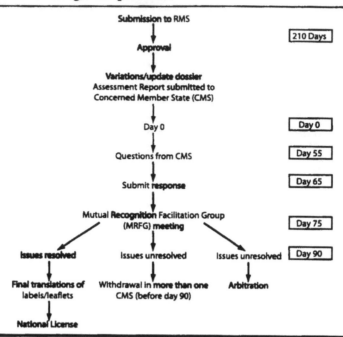

use this procedure. An approval must first be obtained in one Member State and then the applicant can ask other concerned Member States to mutually recognize the approval within ninety days. However, because the initial approval must be updated before the process can begin and the Reference Member State (RMS) must prepare an Assessment Report on this updated file, in practice, the mutual recognition phase can take about eight to nine months after the first approval. Ultimately this procedure produces an opinion and the individual countries must then issue the formal license. This latter activity is very slow in some Member States and remains a concern. About one hundred applications per year, which is double the number using the centralized process, are filed via this process. The Mutual Recognition Procedure does offer more commercial flexibility but, because Member States still carry out their own assessments, the applicant is often forced to answer a large number of detailed questions in a very short time and without the benefit of clock stops.

Both the centralized and decentralized procedures have been under review and there will be further modifications in the future to help improve the operation of both systems.

Local Applications

In the UK, a professional secretariat carries out the review of the Marketing Authorization, but the actual recommendation to approve or reject an application is made by an expert committee made up of academics and hospital physicians. Since decisions are made by the group, individual bias plays less of a role in the final decision. On the other hand, a group makes it much more difficult for a company to negotiate with the EMEA or to gain a clear idea of the reception the application is likely to receive and whether or not there are any problem areas that might be addressed before a decision is made. For this reason, part of the regulatory consultancy role is to gain the support of local opinion from leaders during the early stages of the project, if needed.

A company outside Europe often faces difficulties deciding on which of the various systems available for EU registration is appropriate for the product. If the application is for a single country, then a national application is probably appropriate. However, once a product is intended for more than one country of the EU, the Centralized route or the Mutual Recognition route must be used.

To date, we have been used to three styles of regulatory submissions: those for America, Japan, and Europe. However, an international agreement was reached in November 2000 within the ICH framework on the format and content for applications to be submitted in the three regions. This has led to the Common Technical Document, which it is hoped will save time and resources and, at the same time, assist in regulations review. It has been possible to use this format since June 2001; after June 2003, its use will likely be mandatory, at least for new drugs. Such global dossiers represent a step forward; however, it will be important that these documents are easy to use by the regulatory agencies that have to assess them and that their content remains focused and relevant, rather than a compilation of individual documents. Figure 14.3 is a diagrammatic representation of the ICH Common Technical Document (CTD).

Figure 14.3   ICH Common Technical Document (CTD)

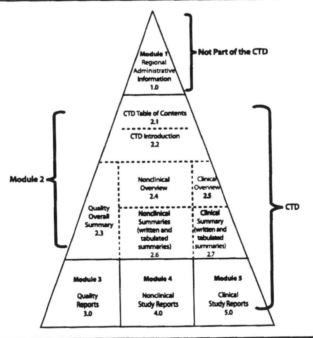

(Adapted with permission from ICH Notice to Applicants 2001.)

WHY USE A REGULATORY CRO AT ALL?

During the development of a product, a large amount of the regulatory work can usually be done much more cheaply by an in-house regulatory person. Why then should a company contract out the work?

Resource

During a project, peaks and troughs of work occur such that the regulatory department will be swamped, then devoid of work. To staff a department to cope with the peaks in order to avoid delaying a project would be economic disaster for the company. On the other hand, to delay the progress of a project because of a lack of regulatory resources would be equally disastrous. Thus, outside assistance at times of high workload is vital.

Expertise

Just as the amount of work will vary, so will the level and area of the expertise needed. Determining the worldwide regulatory strategy requires knowledge and the hands-on experience gained by taking projects through the full development process with all the day-to-day problems and challenges that occur. Expertise will be needed in pharmacology, pharmacokinetics, toxicology, and clinical development, not to mention procedures, appeals, knowledge of the appropriate contacts in regulatory agencies, academia, medical schools, and so on. As the project develops, the expertise required will change and there is no way that a small regulatory department can capture this breadth of knowledge and keep it up-to-date. The CRO, on the other hand, if of a suitable size, will face similar problems day after day and will capture experience that is transferrable to succeeding projects.

Confidentiality

Often, not all information on current regulatory requirements or interpretation is readily available, and it may be necessary to contact regulatory authorities or others to gather the information. A consultant can do this without disclosing the name of the company and, hence, can approach agencies that the company would not feel comfortable approaching directly.

Cost-Effectiveness

Consultants have to provide an estimate of the cost of the work and then stick to the estimate as closely as possible. This makes them well-experienced in assessing the amount of work necessary and its actual cost, something most companies are unable to do. This helps greatly in the decision-making process and leads not only to the optimal progression of projects, but also to the progression of the optimal projects. Very often, companies spend far too much on a project, since they do not know the real internal cost of the people concerned and the overhead they will consume. If the company knew the true cost, perhaps they would not have continued along that particular project path. Companies also fail to take into account the work their employees could have been doing had they not been fully occupied elsewhere.

EVALUATING REGULATORY CONSULTANTS

Scope and Experience

The questions asked of a potential clinical CRO during the selection phase (regarding their global coverage, size, documentation control systems, and therapeutic area experience) may not be so relevant in the regulatory area. In registration, the company's experience in dealing with the specific regulatory agencies involved, their strength of regulatory contacts within the agencies, their geographic representation, and the independent experts available in the clinical, toxicology, and pharmaceutical areas are more important concerns.

Early Interaction

A few years ago, it was possible to hand over documentation to a regulatory consultancy just prior to the scheduled submission date, asking them to assemble the data and make the submission, then to sit back and await the final document. This approach is rapidly disappearing as companies need to shorten the development process by ensuring that data created will meet the expectations of as many regulatory agencies as possible. For this, they need to obtain good-quality regulatory advice and input as early as possible in the project. To achieve the optimum project plan, considerable input, discussion, and interaction are necessary between the company project team and the regulatory consultant. Experience has shown that companies who involve the regulatory CRO at the earliest time and in an open, interactive way are the ones most likely to achieve the highest-quality submissions with the best chances of early success.

Defining the Company Objective

The most aggressive regulatory objective is to gain a license to sell a product in as many markets as possible in the shortest time. A multinational company may have this as their prime objective, but the number of such companies is small and getting smaller each year as more and more mergers take place.

Medium-sized companies may be content with a more modest objective, like gaining approval in a single market or area, recognizing they do not have the resources to cover more of the world. They need to focus on a single country to gain some revenue as soon as possible and prove to potential comarketing partners that their product is registerable. Here the whole development program may be aimed at satisfying a single authority, recognizing that any gaps in the database can be filled in later by a partner.

Smaller companies, such as start-up biotechnology companies, may have an even smaller prime objective. They need to demonstrate that the product is safe and shows some potential activity in order to attract further funding from the business community or from a major pharmaceutical company. In the short term, they need to achieve some defined milestone, such as the first dose to man, or the end of the Phase I work, to be able to continue.

In each case, the regulatory approach will be different and this must be taken into consideration by the consultant when helping to formulate the regulatory plan. Thus, flexibility and experience with different types of companies should be taken into account when selecting the appropriate consultant partner.

CRO Size

Along with increased outsourcing, there is increased competition within the CRO industry, not only in the clinical field, but in the regulatory field as well. Bigger, international CROs now provide a one-stop shop for all pharmaceutical companies' requirements. In deciding whether to go to one of the large organizations or to a collection of smaller specialist companies, the advantages and disadvantages presented in Table 14.1 must be considered.

Table 14.1 demonstrates that nothing is black and white. However, if you wish to hand over the development of a project totally to a CRO, then one of the biggest ones is probably the best choice despite the increased cost. On the other hand, if you wish to be involved with every stage of the operation, overseeing progress and adjusting the process based on emerging data, then a selection of smaller, more specialized companies may be the best approach.

Costs

Clinical studies are so expensive that part of the clinical CRO's objective is to streamline the process to achieve the fastest but most cost-effective trial. In comparison, the costs of employing a regulatory consultancy are much smaller, such that cost itself is much less important than·is commissioning the most appropriate group for the purpose. No one would argue that money spent on designing the trial most acceptable to regulatory authorities is money well spent, since a poorly designed trial means wasted money and the potential delay of product approval.

Consultant prices vary widely, whether the costs are quoted as a per diem or on a fixed-price contract basis. In the regulatory area, fixed prices are a rarity since costs are so difficult to predict in advance, and, in the absence of detailed knowledge of the project and the data available, it is difficult to foretell what problems will arise. Indeed, a fixed-price contract in this field has either been priced so high as to make it an expensive option for the client, or the consultant is so desperate for the work that he or she will take it at any price.

Commercial Viability

With the increased competition in this area, there have been some casualties, with consultancies going bankrupt. Commercial viability should always be examined in the selection of a CRO and cost should never be trimmed to a level that both parties know

Table 14.1   Advantages and Disadvantages of Large CROs

Advantages of Large CROs	Disadvantages of Large CROs
Simplified accounting	Costs likely to be higher
Simplified communications	All eggs in one basket
Good communications between different fields	May be political conflicts within large CRO
Good worldwide coverage	There are probably better local experts if they can be found
Broad band of coverage	Expertise likely to be variable across fields

is noncommercial. There should be some provision in the agreed-upon cost to provide a reasonable profit for the CRO. In this way, the partnership is likely to survive for the duration of the project.

SELECTING THE REGULATORY CONSULTANT

The factors that, in our view, are important in selecting a regulatory consultant are as follows:

- Experience;
- Location;
- Interaction/communication;
- Rapport;
- Language;
- Size; and
- Cost.

Experience

In selecting a CRO, you need to know something of their track record—have they worked through this sort of procedure before and do they have appropriate experts or staff that can work effectively with the opinion leader who is signing the clinical expert report?

The CRO must be fully familiar with the regulatory procedure being considered. It is here that local and current interpretation of the requirements is essential. Do not be surprised, however, if the information you receive from the CRO on previous experience is carefully expressed and lacks detail. This is done to protect the identity of the CRO's clients since many companies insist on strict confidentiality on their activities. The CRO should, however, be able to provide information on:

- Regulatory systems in which they have experience;
- Geographic coverage;
- Curricula vitae of staff;
- Therapeutic areas covered;
- Types of product covered;
- Accessibility to experts on chemistry and pharmacy, toxicology, clinical, Qualified Persons (QP), and so on; and
- Project success rates.

Location

Location is important in the wider sense, since it would be illogical and impractical to use a European-based company to handle FDA matters or a U.S.-based unit to handle European submissions. What is written in the legislation or guidelines or quoted by a speaker in a symposium is not always what will be successful in an application. It is the interpretation of the legislation or guidelines that makes an application successful. An appropriately located CRO is best placed to have a current understanding of the practical aspects of a project and the way to success.

Similarly, in Europe, it is necessary to have local contacts in the key countries and to have constant interaction with the whole range of European organizations involved with drug regulation. Having defined the broad need for location, the more precise location within the specific geographic area is much less important. Thus, the location of the consultant within the United States or Europe in these days of the Internet, fax, and telephone is not a major factor in the choice.

Interaction/Communication/Quality

Having defined the location, the key factor is the degree and quality of the communication between the company and the CRO. Of two similarly experienced CROs, the choice must therefore go to the group with which the company feels most comfortable and with which the company feels that effective communication and a solid working relationship will ensue.

However skilled the CRO, the need for considerable interaction between the groups remains. It may seem at times that the effort required by the company is such that they may as well have done the job themselves. This, however, is a false impression, since by using the CRO, the company is removing most of the compilation work from their own task list, thus freeing up internal resources to manage the project and to properly monitor progress and quality. This cannot be achieved if the company department is struggling to create the application as well. The supervision inevitably leads to discussion on key points of the application and, finally, to a better application overall.

Rapport

All the other positive factors that come from using a CRO are reduced in effectiveness if the pharmaceutical company cannot work productively with the CRO. Defining rapport is not easy since, to be effective, the CRO's advice may not always be what the company wants to hear. It would be a mistake, however, to select an organization that always maintains an argument-free situation and follows the company line. The relationship must be open and there must be mutual respect for the abilities and experience of the parties involved, but this does not mean that they always must agree.

Levels of Contact

One of the major criticisms of some of the larger organizations is that, before the contract is finalized, senior, experienced CRO staff act as sponsor contacts, but once the project is running, junior, inexperienced staff are left to carry out the work. For this reason, the company should always insist on knowing who will be involved in the project. It is unrealistic to imagine that every action in the project will be carried out by the most experienced staff, and the presence of junior CRO members in early meetings may be a good indication that the senior staff intend to maintain involvement and that they are confident in the ability of their junior staff, whom they are happy to introduce to clients. In such a case, the earlier all members of the team meet, the better. The presence of junior staff at early meetings also indicates that the CRO is aiming to create an organization with depth rather than a support organization for a few senior staff, who present the public face of the organization. A support organization lacking depth can work for a time, but must falter as the workload increases.

The objective of the company representative should therefore be to meet as many of the CRO staff as possible during the selection process to answer the following questions:

- Who will actually be doing the work?
- Could I work with these people?
- If the key member of the CRO team was absent, could I still work with the other staff in the short term?

Language

There is no doubt that English is the prime language for European and United States applications; many international companies regard it as the standard scientific language as well. It must be emphasized, however, that European applications are handled by many nonscientific workers who may not speak English fluently. Non-English-speaker involvement, combined with the requirement for labeling in the national language, shows that local language speakers play an important role in ensuring the fastest possible review of an application. This has been well-illustrated by the delays that sometimes occur in obtaining licenses from individual countries after the CPMP has issued a positive opinion in the Mutual Recognition Procedure. Here, locally based, local language speakers can play an important role in advancing the application.

CRO Size

Many large international CROs offer a regulatory service, in addition to covering the clinical and animal safety aspects of drug development. Such a one-stop shop can lead to improved communications between the regulatory group and the clinical and toxicological groups if they all are involved in a project. The downside is that, because it commands a lower contract size, the regulatory group may be dominated by the other groups, in effect acting as a service provider to them, rather than providing independent regulatory advice. Thus, the selection process should include a determination of the level of regulatory group independence and the type and number of regulatory service clients who are not involved with the other groups of the company. Providing regulatory group independence, the advantage of the larger companies is their larger capacity; they can handle bigger projects and are likely to have offices that can provide local support where necessary in a number of countries.

The medium-sized contracting companies tend to be best experienced in local regulatory systems and procedures. They can generally handle the larger projects, but may not have access to the range of experts available to the larger CROs. The choice will thus depend on the match between a company's needs and the specific strengths of the CRO concerned. For long-term relationships including a number of different project types, a larger CRO may well be able to offer a more appropriate service.

Finally, there are the one-man-band operations where the consultant has in-depth knowledge and experience in a specific area and will deal with a project himself without delegating it to a more junior person. If this approach matches a company's requirements, then this choice would be appropriate.

Cost

Accurately estimating the cost of the regulatory function within a project is difficult because a CRO rarely knows the quality of the data or the specific details of the product or of the company's aspirations until after the contract is signed and the CRO has started work. Any CRO experienced in the area would understand, however, that the company needs a cost estimate in order to budget for the forthcoming work and would, therefore, provide the best estimate possible at a very early stage. Because of the nature of the work, a CRO is unlikely to offer a fixed price but it should generate a range of projected costs to slot into the project budget. These projected costs can be refined as the project develops and, with close communications, nasty surprises during the development program should be minimal. Compared with the cost of the clinical work, the registration work (excluding registration fees) will be much more modest.

Because of the high cost, development programs are often split into separate steps with a milestone and decision point between each. This strategy also works for the regulatory aspects of a project and, providing the regulatory group is part of the project team, it is possible to set meaningful milestones and timelines, against which the consultant can be measured. The regulatory submission, however, relies on the availability of documentation from so many parts of a company that the CRO cannot control the timeline. For instance, a regulatory group cannot be judged according to traditional success criteria, such as patient entry rates or clinical report completion.

HOW TO JUDGE A CRO

How can you assess the quality of a CRO? This is a difficult question to answer, because it is similar to selecting or employing staff for a specific project. In both situations, the two parties will be working closely together and the true quality of an employee or a CRO does not become apparent until skills are tested in the project.

First impressions can, however, be helpful in the selection process.

- Does the CRO respond to your first inquiry in a reasonable time?
- Is the quotation produced on time and in the form requested?
- Is the quotation specifically tailored to needs or does it appear pro forma, produced without any specific input?

The quality of the proposal presentation is also important, since it shows the pride the CRO has in its product.

Once the number of potential CROs has been reduced to a workable number (generally less than four), a visit to their offices can provide insight about how they actually perform and the environment and character of the operations.

A company should not expect extensive unpaid work to be performed by the consultants before a contract is finalized. There is a recent trend for companies to try to have their development plan produced by a number of consultants as part of the selection process. If such work is required, it should be paid for whether the contract is ultimately awarded to the CRO or not. This ensures that the appropriate resources are applied to the exercise and that the relationship starts off in a fair and business-like manner.

TASKS IN THE REGULATORY AREA

Regulatory Strategy

At one time, it was possible to develop a product specifically for the originating country, then to extend the development program to allow registration individually in other countries once approval was obtained in the first. Today, the cost of drug development and the pressure on companies to get new products to market as quickly as possible make this approach impossible. It is necessary to pursue parallel registration routes at least in the United States and Europe, since these two markets account for over 60 percent of the total world market for pharmaceuticals. For this reason, a worldwide regulatory strategy is required as early in the project as possible. The strategy document will evolve as the project progresses, but the definition of the company objectives and the steps in the development procedure will remain constant, while the detail and specific direction will change as data becomes available. The act of writing the document at the earliest stage possible clarifies many uncertainties and identifies the key issues for resolution.

Preparation of Applications

There is no doubt that experience yields speed in selecting data and assembling and summarizing the key points for an application. As a project progresses, the CRO is motivated to meet their cost estimates and, more important, their time schedule, since by this late stage in a project, they are looking for the next project and, hence, are driven to perform up to or in excess of client expectation. Repeat business gives the CRO stable income and, hence, long-term survival.

Problem Solving

Few development projects progress without problems. The good CRO can spot problems before they arise or give practical advice, based on experience, on how to resolve problems.

WHEN THE PROJECT IS RUNNING

Once everything is running, it may be tempting for the company to transfer attention to other projects. This can be a mistake! Few projects progress exactly to plan and the need for dialogue and project input is continuous. If this is so, a company may ask the question, why contract out? The answer is that it is a lot easier and less time-consuming to monitor a project than to carry it out and monitor it. Ultimate responsibility for the project rests with the company and this responsibility can never be fully delegated, so constant monitoring and regular, updated financial reviews are wise precautions, however good the CRO. Indeed, the better CRO will welcome the input, since it provides reassurance that the project is on track and that the company has not changed the goalposts. Such input builds trust in the relationship, which can only be generated over time and with input.

Situations where things have gone wrong show that, rather than lack of experience or poor performance, poor communication and lack of clarity in the brief are more often the culprits. For this reason, a clear brief understood by both sides, along with built-in mechanisms of communication and control, is essential from the outset. This requires allocating management time to the project, something often discounted when costs have to be trimmed.

CHANNELS OF COMMUNICATION AND CONTRACTS

Nominating a single contact point within a company and within the CRO can help to ensure that communication is maintained and that both parties understand what the arrangement requires. Of course, each party may interpret the arrangement differently and, therefore, it is essential that the requirements are written down, at least in a letter of understanding and, if the project is of any size, in a formal contract. Because such contracts are pulled together frequently from other contracts, they can prove inappropriate for the regulatory situation that may be:

- Short-term;
- Early in the project, such that only initial ideas exist;
- Subject to continual change as more information becomes available; and/or
- Involving input from a range of specialists.

For this reason the contract should be specific to this type of project and, at the same time, be general enough to avoid constant modification. For this reason, a number of annexes that can be added as the project progresses may be necessary, adding on specific detail as it becomes possible to define. In general, contracts are only read when things go wrong, so from the practical point of view, it is probably more important to have frequent written communications updating the requirements as necessary.

The Present and Future

Like all progressive industries, the pharmaceutical industry is changing. Companies are merging to produce bigger companies. Even the largest multinational companies, however, are now outsourcing more and more work and, in response, CROs too are merging to form larger operations with a wider knowledge base and geographic spread. Thus, the CRO industry is changing also and, although the explosive growth in CROs over the past few years is not expected to continue, the rate will still be very respectable. Like the pharmaceutical industry, there will be fewer and larger CRO companies, but there will still be room for some small niche companies. Regulatory consultancies could operate in some of these niche areas, perhaps encompassing specialist types of products such as biotechnology, herbal, or health food products, where special knowledge and skills are needed.

As the market profile changes, so there is a change in the relationship between the regulatory CRO and, particularly, the largest companies. In the past, the multinationals regarded the CRO as hired hands, selecting CROs mainly on price and availability. Even the preferred vendor arrangements tended to focus on price reduction rather than on a mutually rewarding partnership with shared aims. With small- to medium-sized companies, a partnership approach has worked well for many years, since the success of the CRO activity is central to the success of the company. In effect, they cannot afford to let the CRO fail. It will be interesting to see if the arrangement spreads to the larger companies and whether the shared risk trend in the clinical area will begin to appear in the regulatory area as well.

The expansion of the EU, the increase in ICH activities, and the emergence of the Common Technical Document will change the approach to registration, but it is unlikely that the three major types of agencies (FDA, EU, and Japan) will converge entirely in their needs or in their interpretation of requirements. There will, therefore, still be the need for local specialist regulatory CROs to provide the on-the-ground advice necessary for a successful application.

Recently, a further hurdle to getting drugs to the market has emerged—the move to limit the list of drugs that are reimbursed by national governments. A new set of skills is necessary for overcoming this obstacle; specialist CROs will emerge to fill this gap.

CONCLUSION

Not all CRO activities are the same. In the clinical field, the aim must be to select the most efficient and cost-effective operator. For the bigger trials, this is likely to be one of the bigger organizations with global coverage and well-established systems and procedures. In the regulatory field, this may not be so. In regulatory work, a close and respected association must develop with a group of people who are experienced in the specific geographic areas involved, who can relate productively to the company, and who are flexible enough to take account of specific product problems and opportunities, rather than applying rigid systems and procedures.

This kind of relationship takes time to develop; the earlier the registration group becomes involved with the company, the better. This may not always be a company's least expensive option, but the rewards are large (in terms of shortening the time to license approval) and the expense is likely to be insignificant in the total project cost (in comparison with the cost of a clinical study or, particularly, a wasted clinical study).

REFERENCE

European Commission. 1998. *The rules governing medicinal products in the European Union.* Vol. 2A, *Notice to applicants: Medicinal products for human use; procedures for marketing authorisation.* Vol. 2B, *Notice to applicants: Presentation and content of the dossier.* Brussels: Office for Official Publications of the European Communities.

Jeffrey S. Rudolph and Joseph S. Tempio

Trends in Pharmaceutical Development Outsourcing

Although the pharmaceutical industry is a leader in scientific and technical innovation, it has been slow to adopt business enhancement practices successfully implemented years earlier in the automotive, electronic, and aerospace industries. In response to business pressures or crises, other industries have gone back to the basics and have progressed in line with a fundamental business analysis where core competences—those skills that provide differentiation and therefore competitive advantage—are identified and enhanced through internal development and external access. Those activities identified as noncore activities are contracted out, leaving personnel and financial resources focused on the highest value-added business components.

Although the pharmaceutical industry is not yet in crisis, there certainly are mounting pressures already felt on the bottom line and in the boardrooms. Escalating R&D costs, extended development times, limited patent life, generic competition, and reduced profitability, resulting from managed care formularies and price controls, are just a few of the warning signs that crisis may be just around the corner.

One reason the pharmaceutical R&D community has been slow to react to these pressures is that the traditional organizational and management structure has been based on departmental or functional silos where management have been rewarded on the basis of their success against specific functional, as opposed to company, objectives. Some of these managers, seeing the potential for internal staff reductions, or at least a slowing of staff growth because of outsourcing initiatives, have become concerned about their departments' stature within the organization. They are aware that they lack the skills to do what needs to be done and, as a result, they are being forced to access external skills. Additionally, department headcount in many firms is an indirect measure of the importance of a functional area and the fiefdom would be in jeopardy should outsourcing be actively pursued as a business strategy.

BACKGROUND

The roots of pharmaceutical development outsourcing growth over the last two decades lie in the convergence of political, economic, and philosophical factors. These factors are discussed in depth in the various chapters of this book. But, it is important that we discuss to some extent the dynamics of those changes, so that we may better be prepared to identify the forces that may shape the next few decades.

In the early 1980s, recession and the increasingly high cost of drug discovery and development forced many pharmaceutical companies to cut expenses. The industry as a whole was also driven to evaluate consolidation, though in an industry where mergers had been traditionally friendly, the pace toward consolidation was slow. The threat of healthcare reform in the United States at the beginning of the Clinton administration, the inroads of generic competition, and the continuing increase in regulatory requirements for drug development worldwide placed continued economic pressure on pharmaceutical companies to cut costs. In addition, the retirement and healthcare costs for workers in the industry, who were in general more highly compensated than in other industries and living longer, fortified the economic pressure for the industry to eliminate jobs or shift employee costs to others.

The threat of healthcare reform and its inevitable impact on profit margins seemed to be the catalyst for bold action. In a short time, pharmaceutical industry executives developed two distinct philosophical concepts of the definition of a drug company. One school of thought seemed to hold that drug companies needed to position themselves as full-service providers (bundlers) of the healthcare commodity. Supporters of this concept thought that pharmaceutical companies could most profitably and cost-effectively deal with the large healthcare providers and intermediaries—governments and government-sponsored buying groups, insurance companies, and healthcare providers like Healthcare Management Organizations (HMOs)—that would be their principal customers by offering these companies a cafeteria of bundled healthcare products. These companies proceeded to acquire lower-margin drug distribution providers and generic and hospital supply resources to support that industry concept. These companies were positioning themselves to compete for market share by offering a full range of products to the large institutional buyers, believed to be the future dominant customers. At the same time, these companies continued to invest in R&D, since new products would command premium prices and have exclusivity for a time.

The alternative vision was that drug companies were in essence drug discovery enterprises, research companies that also had exceptional expertise, and communication, and distribution resources for the marketing and sale of new drug discoveries. These companies concentrated resources in developing their R&D pipelines. They reasoned that profit margins, in the reformed healthcare environment described above, could be preserved best by bringing new drugs to the market at premium prices and high profit margins.

Both drug company concepts required that the major pharmaceutical companies accept the reality of diminishing price flexibility and declining share for older drugs that had lost patent protection. This would be due to the increased competition from generic drugs, and the bargaining clout of the institutional buyers.

In either case, it was also necessary that the cost structure of the pharmaceutical companies be reduced to provide the capital required for increased R&D investment. People costs, particularly healthcare and retirement costs for higher-seniority workers, were major expenses for the industry. Therefore staff reduction was a point of focus. A concerted effort was made to make the workforce smaller and younger. Transfer of personnel costs to a lower compensation structure in a service provider and relief from the attendant retirement liabilities were expected to cut direct expenses. In addition, outsource resources were flexible; they could be used on demand and discontinued at will. These factors made tactical outsourcing attractive.

Drug development services were available to some extent at that time from niche development resource suppliers. Demand for development resources tended to rise and fall with the richness of the demands of the R&D pipeline and in the "drug discovery paradigm," there were no core resources. Thus, some companies maintained a reasonable baseline development staff while peak demands were outsourced. In some cases, downsized workers were hired back as consultants to complete programs in progress; in other cases, clinical contract service organizations (Contract Research Organizations, called CROs) or contract service providers were used.

Contemporaneously with these changes in the pharmaceutical industry, the United States biotech industry began to bring its first products through development, and new high-tech and biotech start-ups were proliferating. Since these companies sprang from a technological base different from the traditional small molecule pharmaceutical industry, the core technical staff was often naive about the drug development regulatory process. Because biotech start-up companies lacked the depth of development and regulatory infrastructure necessary to develop products through Food and Drug Administration (FDA)/government licensing, they required outsource resources to complete commercialization of their innovations.

Many major pharmaceutical companies downsized their development organizations, primarily through the early retirement of workers and through controlling the growth in development headcount. In addition, mergers and consolidations resulted in the dislocation of relatively large numbers of drug development personnel, since these skills were most apparently redundant in the merged organization. The resulting staff reductions had several effects on the pharmaceutical and the outsourcing industries. For companies with the richest R&D pipelines, the reductions in the development workforce made it necessary to outsource drug development work tactically. The peaks and valleys of pharmaceutical development often made it necessary to smooth peak demand for development resources with outsourcing. Thus, the leading companies in R&D were outsourcing, giving credibility to the strategy.

Highly educated and skilled development personnel recently separated from pharmaceutical companies entered the outsourcing services industry, either as employees or entrepreneurs. This bolstered the quality and quantity of outsourcing resources. In addition, some highly educated and skilled pharmaceutical industry personnel entered the pharmaceutical industry to start up new pharmaceutical companies, while others entered the biotech industry, often in start-ups. In general, this tended to increase the demand for outsource development services, since internal development resources were often scarce or nonexistent in start-up companies. And finally, consolidated companies frequently did not retain sufficient resources in the clinical research and development and clinical support areas (clinical supplies manufacturing, analytical support and quality control, clinical packaging, and so on) to support the merged development pipeline peak demands. These companies often turned to outsourcing to remedy these transient shortages.

In essence, the economic and political pressures on pharmaceutical and biotech companies and the growth of the biotech industry tended to fuel the growth and development of the outsourcing industry. The impact was felt in two ways: by increasing demand for outsourcing services and by increasing the pool of experienced and highly competent workers available to the development services industry. In addition, the biotech and pharmaceutical industries had a vested interest in the success of outsourcing, since outsourcing was seen as both a viable tactical strategy to contain costs and a potential way of decreasing development cycle times by eliminating priority conflicts and scheduling delays. Therefore, pharmaceutical companies made a concerted effort to monitor their contractors' compliance and, in some cases, invested in the capabilities of the service providers. Frequently, the demands of pharmaceutical companies resulted in the broadening of services offered by CROs.

The preceding factors have contributed to the explosive growth of the drug development services industry over the last two decades. As the industry developed, niche suppliers broadened offerings or merged with others to the point where several development service organizations claimed the ability to deliver a full array of development services internationally. The original emphasis of development service organizations on experience, customer service, and speed has subtly evolved to a point where the industry leader has mirrored the development and support services within a major ethical drug company. Examples include training, validation services, and marketing and sales capability. These service offerings include discovery technology offerings such as combinatorial chemistry, as well as full clinical development services, prelaunch technical information development, and marketing and sales services. Therefore, with the exception of products, the available infrastructure in the largest development service organizations mirrors that of a pharmaceutical company.

The strategic focus of the outsource industry leader appears to have changed from a service company to a solutions company—a provider of expertise. Some believe that the vision of the development services industry market leaders is that in the future, development service organizations, not pharmaceutical companies, will be the experts in drug development. This vision appears to be at odds with the viewpoint of the pharmaceutical industry R&D executives, who have mostly used outsource resources tactically to fill resource gaps. Moreover, some industry analysts believe that these fully integrated companies will ultimately divest those services that are not core to the delivery of traditional preclinical and clinical contract services. However, the vision of the integrated full-service development services organization is consistent with the needs of a drug company configured as a discovery/marketing/sales enterprise.

However, at the present time, most large pharmaceutical companies use development service providers tactically, to fill resource gaps, and even in preferred provider relationships there generally is no real integration of resources. Though there have been some attempts at strategic integration of in-house and outsource R&D resources, and some situations where outsource companies were given a New Chemical Entity (NCE) to develop virtually alone, these are exceptions. For outsourcing to reach its full potential, a collaborative structure and appropriate supporting managerial structures must emerge. In addition, pharmaceutical companies in general must accept service providers' leadership in those areas deemed not to be core competencies. This will require a change in mindset for pharmaceutical companies. The following section will explore some pressures that may drive these changes and shape the future of drug development outsourcing.

FUTURE DRIVERS

Political Factors

As mentioned previously, the healthcare reform debate in the United States early in the Clinton administration caught the attention of the pharmaceutical industry and resulted in bold action. Several major pharmaceutical companies acquired distribution, generic, or pharmacy benefit management resources in order to position themselves to cope with a new paradigm where the major purchasers of drugs would be governmental agencies, insurance companies, institutions, or their surrogates. With the failure of these healthcare reform initiatives, and no revolutionary change in the way drugs were purchased, many of these assets were quietly absorbed or divested by their drug company owners. But, recent political events in the United States may revive interest in this model. It is too early in the G. W. Bush administration cycle to predict how it will impact healthcare expenditure, but with Congress so evenly divided, it is not likely that any changes will be revolutionary or even far-reaching.

There is significant consumer awareness of the full cost of prescription drugs in the United States, since many consumers are still directly involved in this transaction. Political pressure for cost controls on prescription drugs has been newsworthy and activists often target drug prices both nationally and internationally. For instance, there was significant adverse press on the cost of drugs for treatment of impotence. Also, there has been almost constant controversy over the price of AIDS/HIV drugs in the United States, and the availability of these drugs in southern Africa in particular. Recent actions by the South African government to allow local manufacturers to violate AIDS drug patents are notable, as is the eventual acquiescence of the drug companies to this policy. Furthermore, several state governments of the United States have raised the issue of the cost of prescription drugs in the United States compared with the cost of the same drugs in Canada. These price differentials are significant, as much as two- to threefold at the retail level. This is a major issue among the consumer activist groups, but even in 2001, they were no nearer to resolution. Thus the industry can expect continued consumer and political pressure on drug prices in the future from both state and local governments, and from ambitious politicians.

Demographics/Social Factors

The baby boomers of the post-WWII period are into their fifties. Although the global population is becoming younger on average, in the Organization for Economic Co-operation and Development (OECD) countries the reverse is true. While an average of 9 percent of the population was sixty-five or older in 1990, it is forecast that by 2040, over 22 percent will be of that age. Many new drugs introduced to the market in recent years treat the diseases or enhance the senior citizen's quality of life. Popular interest in wellness—the prevention of aging and the preservation of quality of life into old age—is growing. It is also likely that this generation will have substantial means, and if the turmoil of the 1960s is any indicator, a willingness to take control of their medical care and embrace nontraditional solutions. One may conclude from these social and demographic factors that there will be a high demand for new drugs, a sense of urgency about bringing new products to patients, a willingness to experiment with new therapies and nontraditional therapies, and a demand for expedited new drug licensing. (Recent concern over the withdrawal of a number of recently licensed drugs for toxicity reasons has, however, prompted the question of whether the FDA has tilted the balance of speed versus rigor too far on the speed side.) In addition, there will be substantial and increasing financial demands on the healthcare delivery system.

These social and demographic pressures are sure to drive political tensions between the healthcare requirements of the retiring baby boom generation and the infrastructure, education, employment, and security requirements of Generation X. It is likely that these tensions will create profit margin pressures on pharmaceutical companies, as politicians strain to balance public priorities.

The current process of registering and licensing drugs, evolving as a result of International Conference on Harmonization (ICH) and other initiatives, is sure to come under scrutiny as the financial pressures on drug companies and the unfulfilled therapeutic needs of consumers clash with the high cost and long delays in the current drug approval process. It is unlikely that consumers will allow regulatory authorities to deny patients access to important drugs based only on national sovereignty or geography. Harmonization of regulatory requirements will become a populist political issue. Information technology and the Internet will make information about drug discoveries and the progress of drug innovations worldwide available to all interested parties. The medical establishment and regulators are sure to feel pressure from well-informed consumers to make innovations available simultaneously worldwide. Perhaps information technology and the Internet may develop as formidable tools in reaching a goal of rolling regulatory reviews and expeditious approvals.

Internet/Information Technology

The Internet and the information revolution, which is having far-reaching effects on all aspects of daily life, will likely strongly impact the drug development and regulatory process. Real-time communications capability makes it far less necessary for colleagues working on the same project to be in the same venue. Also, information technology enriches the potential for innovation by making real-time data and more complete data analysis available to more collaborators. In addition, it provides the potential for better drug design, screening, selection, and the promise of simulation of in vivo performance. In fact, in the future, it may be possible to substantially accelerate drug development by eliminating all but the most pivotal of clinical trials, the preceding work having been simulated from genomic, pharmacologic, pharmacokinetic, and biopharmaceutic trials. Coupled with outsourcing, information technology and the new discovery technologies may also serve to lower the barrier to entry into the pharmaceutical industry, perhaps shaking the hold of the major multinational drug companies.

The pharmaceutical industry has been relatively slow in adopting new technology for clinical information system management. Because clinical data management is one of the clinical CRO's core competences and because the industry is relatively young, clinical CROs often have newer and better technology than the pharmaceutical clients they serve. In addition, these companies have the ability to evaluate and adopt best practice and to perform many more trials than a single company. They also have extensive experience in study design. This gap in technology and experience may eventually lead to a decision by most pharmaceutical companies to forego further modernization and investment in clinical data management infrastructure in favor of near total outsourcing of this function to CROs. There are significant opportunities for gaining time in drug development associated with the expeditious handling of the clinical study data. The time between the point the last patient is completed and the study report is ready to file is a time increment that can be significantly shortened by appropriate technology and design. Several CROs claim prowess in this area.

Information technology can also substantially impact the compilation and review of regulatory filings and their transmission to regulators throughout the world. Regulatory agencies are currently coping with standard format and data requirement issues for new drug approval. Much progress has been and is currently being made and, although a format for the Common Technical Document has been agreed upon in principle by the parties to ICH, this document has yet to be implemented in an electronic format. However, the FDA has mandated that electronic submissions only will be accepted after 2002, although the FDA has made similar decisions before and privately admits this deadline is likely to be pushed back.

Electronic media can facilitate the expeditious presentation and evaluation of different presentations of the various permutations and combinations of clinical findings across studies and patient subsets within and across studies. The timely availability of this type of data and analysis to the regulatory reviewer can lead to a more expeditious and accurate evaluation and, potentially, to quicker action.

Many CROs have been able to post sponsor study data on the Internet in real time for a number of years and make it available to all concerned with the progress of a clinical trial. It is likely that in the future all parties, including investigators, regulators, sponsors, and study coordinators, will have free access to regulatory data in real time. This would allow for the regulatory authorities to conduct a rolling, real-time review, and perhaps to cut approval times further. The obstacle to this initiative being more widely adopted seems to be the plethora of competing, incompatible systems, few, if any, of which seem to have any advantage over the others. In addition, there is the promise of real-time data collection and analysis of Phase IV and other relevant surveillance data, which could be widely available to mediate the risk of expedited approvals.

Economic Factors

The cost of bringing a new drug through the regulatory process and to market has been covered previously. These costs have risen steadily over the last two decades. Still, as drug companies have merged and consolidated, the R&D investment required to bring enough new drugs to market to support the traditionally high rate of return and growth rate of the pharmaceutical industry has risen, to a point, but with diminishing returns.

Today, pharmaceutical firms are reinvesting between 17 and 20 percent of their sales back into R&D. The Pharmaceutical Research and Manufacturers of America (PhRMA) estimates in its 2001/2 annual report that United States pharmaceutical companies will have spent more than $30 billion on research and development in 2001, an increase of almost 19 percent over 2000 (PhRMA 2000). Globally, the total figure will approach $60 billion if biotechnology companies are included. By comparison, in 1978, reinvestment was only 10 percent of sales. Although R&D expenses are growing at a double-digit rate, reflecting the high costs of new chemistry and biology techniques, equipment, and so forth, as well as the near doubling of the regulatory requirements for clinical studies since 1987, company revenue has grown at only some 7 percent and is not expected, without major changes, to exceed this figure through 2005. This economic gap, coupled with the predicted loss of current product revenue over time, will create a shortfall in company profits, jeopardizing the economic foundation of pharmaceutical firms for the future and reemphasizing the importance of new, commercially successful products (see Figure 15.1).

These economics suggest that R&D must reinvent itself with respect to the way it discovers and develops drugs. The new chemistry, biology, and genomic revolution will, in time, create a richer pool of innovative, high-quality drug candidates. Discovery management believes that its function will be able to triple the number of development-ready drug candidates within only three years. However, with this greater discovery productivity come the downstream implications of a greater number of compounds entering development, potentially forcing many of these high potential leads to go undeveloped. This situation, coupled with limited manpower growth associated with fixed cost constraints as a critical component of maintaining profitability, will result in

Figure 15.1   New products required to sustain growth

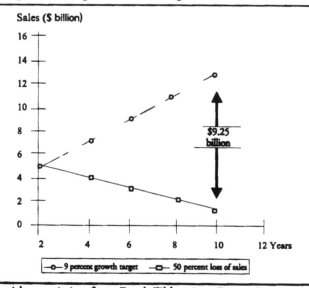

(Figure reprinted with permission from Frank Fildes, AstraZeneca.)

companies experiencing significant bottlenecks in the downstream development-for-launch processes. Do these situations create an opportunity for an increased level of outsourcing, both tactically and strategically? Figure 15.2 demonstrates how new R&D techniques, coupled with increasing outsourcing, produce substantially more registered products than traditional R&D methods.

Analyses by CMR International (2000) suggest that, on average for the next five years, the industry will launch new products based on 1.8 new active substances not hitherto marketed—far short of the three to six drugs predicted to be necessary to ensure reasonable business growth, as shown in Figure 15.3 (Ogg et al. 2000). A recent report from Accenture (2000) further emphasizes the need for the major companies in particular to accelerate their drug development programs and to make them more efficient.

OUTSOURCING INDUSTRY—MATURATION

The drug development outsourcing industry is tiny and immature when compared to other such service industries, for example the service industries that support automobiles and electronics. Few companies have been in business for two decades, and until recently, few had the financial stability and stature to execute global development programs and to support large, megatrial clinical development programs. In the last several years, after a period of explosive 30 to 60 percent industry growth, and with access to capital markets, the industry has entered a phase of sustained growth and industry consolidation. The latter is driven by a need for mass, scale, and global reach that has forced companies to broaden services, increase capacity, globalize, and raise large amounts of capital to fund the expansion. Industry watchers believe that over the next several years development service companies will be forced strategically to situate

Figure 15.2   R&D—future outsourcing opportunities

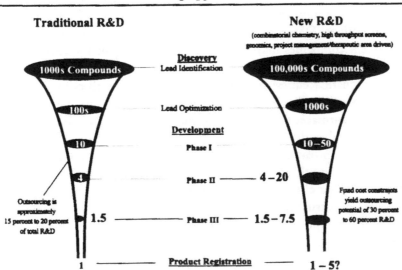

Figure 15.3   Industry R&D targets—to sustain growth at 10 percent

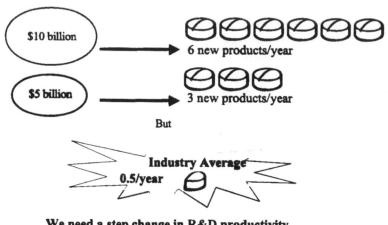

But

Industry Average
0.5/year

We need a step change in R&D productivity

(Data compiled from Ogg et al. 2000.)

themselves as a niche player or to become or merge with a megadevelopment service organization or related corporate entity, for instance a pharmaceutical company or large service provider.

The most recent growth rates of companies in drug development services support this analysis, in that the industry leaders, with few exceptions, have enjoyed the fastest growth rates. This creates a survival problem for the midtier companies who must grow to be more attractive to sponsors. It is generally believed that within five years most midsize organizations will all but disappear. This will largely occur as a result of industry consolidation through mergers.

The top-tier companies in the drug development services industry have positioned themselves as full-service, comprehensive suppliers of all the necessary services required to discover, develop, and launch to market new drugs and devices worldwide. These companies have accomplished this largely through acquisition of other Development Service Organizations (DSOs) or related service companies. This has been driven by a vision that client companies may demand services that span the entire product life cycle. These companies are well-positioned to satisfy the tactical requirements of the large and small pharmaceutical companies and are equally well-positioned to partner or enter into a strategic alliance with drug companies of any size. They are in essence fully integrated pharmaceutical service organizations that, apart from products, are functionally identical to drug companies.

As suppliers to a broad client base, these enterprises are in an excellent position to evaluate best practice throughout the pharmaceutical industry and to leverage this knowledge across their client base. Over time, these companies have the potential to build expertise, infrastructure, and systems that are superior to any individual client. In addition, the composite information available to these companies across the industry with respect to patient populations, safety and efficacy of individual compounds, outcomes research, and study design can be leveraged to improve and accelerate clinical development and drive innovation. The richness of these information assets may well exceed the value of a simple service provider and be an important factor in producing closer integration between client pharmaceutical companies and the outsource industry.

In terms of future demand, CROs will find their resources and technology services across the breadth of preclinical and clinical arenas sought after to an even greater extent than they are today. Table 15.1 shows that CRO use in all R&D areas is predicted to increase significantly as a result of R&D outsourcing demands.

Will the CRO industry be able to cope with these increased demands? Will they have the skills, manpower, and infrastructure to deal with these opportunities? They are undergoing their own revolution in terms of consolidation to create firms of larger critical mass and breadth of vendor offerings to attract clients who are interested in one-stop-shopping. Also, niche players continue to find an attractive marketplace among clients who are interested in unique technologies or who just desire a more personal touch and possibly a faster turnaround, which can be provided by these smaller, less bureaucratic service providers.

Consolidation/Globalization

It is very likely that the trend toward globalization and consolidation in the pharmaceutical industry will continue. The economic forces behind this have been previously discussed. These same factors will, in large measure, force consolidation in the service sector. It is not unthinkable that hybrid mergers will also develop as the pharmaceutical industry reconfigures itself to assimilate the appropriate elements of biotech, service resources, distribution resources, and the like. It is likely that the niche players from the shakeout of these consolidation activities must have regional, technological, or service advantages to survive.

DISCOVERY—OUTSOURCING WITH A TWIST AND FLAIR!

Pharmaceutical companies have renewed their commitment to innovation-focused R&D as a primary producer of growth. The Accenture survey (2000) reports that executives expect to increase the number of compounds entering clinical development by

Table 15.1 Estimated Sizes of the Outsourcing Markets ($ billion U.S.)

Functional Area	Market Size 1997	Market Size 2002	Percent Increase (1997–2002)
Clinical	2.9	5.2	79
Research	1.3	1.8	38
Chemistry, manufacturing, and controls	0.9	1.4	68
Regulatory pharmacology and toxicology	0.5	0.9	80
Total	5.6	9.4	68

(Hughes and Lumley 1999. Reprinted with permission from Technomark Consulting Services Ltd.)

65 percent by 2008, to improve success rates from about one in ten to nearly three in ten, and to reduce times from first dose in humans to regulatory approval by 33 percent. Discovery management plans to deliver higher-quality, "development-ready" compounds with greater innovation to increase likely sales potential as well as to lower the development attrition rate. This is important because it is still estimated that only one in three drugs that go onto the market make a return on their investment. It is far from obvious that these objectives can be achieved with current resources or by using the current drug development paradigm. The CMRI annual report for 2000 indicates that there has been little progress in speeding up drug development or in making it more efficient (CMR International 2000).

Research executives report that innovation, the powerhouse to fuel company growth, has not improved in recent years and that, in addition to implementing the right project/portfolio strategy within a high-performance organization, strategic alliances are fundamental to achieve greater leverage and innovation. Technology alliances, rather than capacity management, are what drive outsourcing within discovery. Figure 15.4 summarizes the key issues necessary to achieve leverage and to capitalize on outside innovations.

Alliance deals with external collaborators tend to run from $20,000 to $1 million with individual academics to $1 million to $100 million with large companies. Most of these are in tools, techniques, and full-fledged research partnerships to find better targets and to execute faster.

The biotech industry remains highly attractive to pharmaceutical companies with partnering rising sharply. This type of relationship will continue to rise in popularity in the future to deliver the unmet needs of the pharmaceutical discovery function. However, pharmaceutical partnering with academia will flourish perhaps at a greater rate because of its focus not only on cutting-edge, innovative research tools, techniques, and discovery leads in specific therapeutic areas, but also because collaborations will yield publications that will benefit the entire research community.

The management of these collaborations—obtaining a balance between internal efforts and achieving maximum value from all resources—is no simple task. Some firms provide significant effort to support the external alliances, even at the expense of internal project management. However, it has been proven by many examples that companies that fail to apply management commitment and expertise to external alliances will risk disappointment, jeopardize milestone deliverables, and increase the level of overall frustration with the organization.

In order to help ensure successful external collaborations, the following tools will assist the sponsor company with managing the collaboration:

- Establish an alliance strategy based on strategic choices about in-house core competences that provide competitive advantage and the availability of qualified and motivated alliance partners with whom to form successful collaborations;
- Devote specialized (project management, financial, purchasing) people in-house to the collaboration;
- Conduct training in relationship management;
- Set clear goals, objectives, and timelines up front;

Figure 15.4   Key issues necessary to achieve leverage and capitalize on outside innovations

- Collaborations make up an increasingly important component of companies' strategic mixes, with a trend toward **20 percent of the discovery budget going to external alliances** over the next few years.

- While biotech and **university-based partners** are the main beneficiaries of this trend toward increased externalization of research expenditures, **small private companies** are also to fill the gaps.

- The growing popularity of collaborations brings with it its own set of challenges, as companies try to figure out **what to outsource and at what price.**

- As companies devote increasing resources to collaborations, they universally struggle with the additional challenge of *managing* these **collaborations** to maintain a balance with internal efforts and achieve maximum value.

- Establish frequent, open communication; and
- Consider offering technical and managerial training programs to the supplier to supplement their own knowledge.

DEVELOPMENT—COPING WITH QUALITY AND QUANTITY

The development arena, preclinical and clinical, is currently faced with sourcing issues that relate to meeting project deliverable timeline objectives, but within fixed cost constraints. Outsourcing has progressed using a tactical approach mainly, although several clinical departments have progressed strategically into preferred provider relationships. Figure 15.5 depicts the supplier relationship continuum.

Achieving the desired quality within the timeline boundaries remains the prominent issue. As the CRO industry reorganizes into mainly large, full-service, potentially bureaucratic organizations, they find delivering projects in a timely, efficient manner difficult, similar to the difficulties faced by major, large pharmaceutical firms. Thus, some clients may opt to use smaller niche organizations that can provide the personal attention that client companies originally sought when they began looking for external resource collaborations.

With the potential increase in development workload associated with the future enhanced output from discovery, client companies will find that strategic sourcing is the only way to meet project objectives. CROs will be in even greater demand, preferred provider or partnership alliances will increase, and, in time, client companies that are slow to react to these changes will find few CROs remaining to establish long-term relationships with. Will pharmaceutical companies vertically integrate as they did with pharmaceutical benefit management companies some years ago (although now they are breaking those linkages) or will there be another paradigm? One thing is sure, the dynamics will change constantly and both the client and supplier companies must think strategically and retain maximum flexibility if they are going to achieve their overall business objectives.

EXTERNAL ENVIRONMENT NONPHARMACEUTICAL OUTSOURCING/BEST PRACTICE

There has been increasing evidence from a variety of industries, especially electronics and engineering, that strategic outsourcing can make a very significant contribution to a business's competitive position, financial performance, and even survival.

The pharmaceutical industry has to address a number of critical challenges, especially with respect to R&D productivity, shorter time to market, and the adoption of the so-called new R&D paradigm. There is little doubt that outsourcing or subcontracting will play a greater role in the strategies adopted to meet these challenges.

In December 1997, InterMatrix and Technomark Consulting Services jointly launched a consortium project. The prime objective of the project was to identify and interpret best practices in strategic outsourcing in nonpharmaceutical industries for the sponsors who were both pharmaceutical companies and key suppliers to R&D. The intent was, through the benchmarking of best practices, for the sponsors to learn how to adopt the knowledge of others to benefit the development of their own procurement and sourcing strategies, as relevant.

Figure 15.5   Supplier relationship continuum

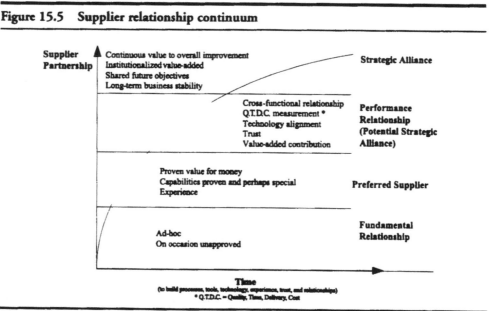

(Figure by Nadia Turner and Chris Keep. Printed with permission from AstraZeneca.)

Eight companies were targeted for benchmarking. Each was a well-respected company and each had a proven track record in relation to strategic outsourcing—as a customer or supplier. Table 15.2 lists the benchmarked companies and their relevant characteristics.

Although the motivations, types of business, and characteristics differed greatly between the benchmarked companies, their policies, approaches, and prime operational factors have had a high degree of similarity between them.

Survey Results, Conclusions, and Implications

1. Definitions

Outsourcing strategy is the explicit policy, with related implementation requirements, on what the business will carry out itself and what will be entrusted to a third party.

A *best practice outsourcing strategy* will identify the different types of relationships with suppliers, generally conditioned on the risks related to the activity, its commercial significance (i.e., its ability to affect profit), and the structure of the supply sector (e.g., many or few specialist suppliers). It will also be based on clearly defined core competences.

Strategic outsourcing is confined to the small number (relative to the business) of key activities in which significant value can be delivered by the supplier, facilitated by having a partnership-type relationship. "Significant" is defined by the volume of business linked to the relationship.

Strategic outsourcing redefines the relationship between the customer and the supplier, in that the new relationship is based on entrusting a complete activity (or set of activities) to the supplier. Strategic outsourcing may imply sole sourcing for the activity. Strategic outsourcing is *not* the improved management of tactical outsourcing.

Two types of strategic outsourcing may be considered. The first, and more common, is designed to enhance the capacity of the business through what is, in effect, virtual vertical integration. This model is now described as the *Extended Enterprise*. The key aspect of this model is that strategic outsourcing does not directly broaden the company's portfolio of products and services. The second type is designed to expand the company's offering with partnerships, which add technology, products, and services. These are *horizontal alliances*.

The Best Practice Model tends to focus on the Extended Enterprise, but not exclusively.

2. Strategic development and framework

The benchmarked companies demonstrated clear strategic vision of the shape and future direction of their organizations, together with top management support and visible commitment to outsourcing as a means of realizing the vision and objectives. This included the ability to remove any endemic resistance to change.

Key features of the outsourcing strategy include:

- A clear understanding of value-creating activities and core competences;
- An organization and activities that are customer-focused;
- An established process for making strategic outsourcing decisions; and
- Senior management having a mandate for outsourcing policies and practices.

Table 15.2　Benchmarked Companies and Their Characteristics

Company	Business	Country	Drivers for Strategic Outsourcing	Other Factors
Company A	Stockholder services	U.S.	Technology acquisition Resources	Strength in defining customer requirements
Company B	Aerospace regional jets	UK	Survival	Ten years' experience
Company C	Airline	UK	Shareholder value Focus on core competence	Model for subsequent BA initiatives
Company D	Automotive	U.S.	Survival Competitive differentiation	Viewed as a leading U.S. best practice
Company E	Research center	Italy	Resource management Technology acquisition	R&D focus
Company F	Fiber-optic cables	UK	Customer acquisition	Won 100% of market in four years through strategic supplier positioning
Company G	Software development	U.S.	Technology/skills acquisition and capacity development	Strong process-driven approach
Company H	Test equipment	UK	Resource management Value-added efficiency	Core to prosperity as IBM MBO

3. Internal organization for strategic outsourcing

One of the most complex aspects in implementing strategic outsourcing in R&D is the separation of the respective management of:

- The R&D process, including the outsourcing process;
- The supply relationships; and
- Specific research and product development projects.

One consequence is that outsourcing decisions are shared responsibilities. Thus, the decision about which supplier to partner with is made by process managers and representatives from various function departments; decisions about which projects to outsource are made by the project team and representatives from various function departments.

Overall relationships with individual strategic suppliers are managed by multifunctional teams, frequently led by purchasing. The benchmarked companies all recognized that the effective management of strategic suppliers requires particular commercial skills.

A key feature of the internal organization for strategic outsourcing is the evolution toward an enterprise model, whereby the strategic supplier is aligned to customer objectives, with rewards linked to the customer's success. The supplier becomes part of the (extended) enterprise.

4. Decision and selection

Strategic decisions on outsourcing are taken by senior functional and operations management independently of specific suppliers. Decisions are based on an explicit evaluation of known or desired core competences/technologies, resource constraints, and/or the balance of risk and reward.

The selection criteria for strategic suppliers include strategic alignment, cultural fit, similarities/complementarities of processes and practices, as well as "quality, cost, and delivery." The benchmarked companies put great emphasis on strategic alignment and cultural fit.

5. Managing the strategic relationship

Strategic supplier relationships are characterized by multilayer, multifunctional communications, although these relationships are managed according to defined parameters and responsibilities. Clearly, this is only justified if the value of the relationship is significant.

In other respects, the relationship is managed based on:

- Mutual trust;
- The minimization of detailed legal documentation;
- The expectation of increasing value over time to the customer and value added from the supplier;
- Early involvement of the supplier in projects; and
- A continuous improvement in focus with an enhanced client-supplier relationship.

With few exceptions, relationships were managed without formal contracts for individual transactions—even among United States companies.

Almost all the benchmarked companies had developed data or adopted approaches so that pricing has become relatively transparent, removing price and related negotiation as a source of conflict.

6. Implications to the pharmaceutical industry

Effective strategic outsourcing will require major changes in the way pharmaceutical companies and their suppliers operate. Value management linked to effective risk management will be core to the process. Major changes can be justified only if the expected value, in all respects, sufficiently exceeds the costs of implementation. Fortunately, the gains for the benchmarked companies have been substantial; in other words, strategic outsourcing pays.

Pharmaceutical companies will need to define clearly their core competences in R&D before identifying the areas most appropriate to strategic outsourcing. The approach is unlikely to be fully suitable for capacity management, unless it applies to a complete development program. Outsourcing management is likely to become a core competence.

The implementation of strategic outsourcing can take years to embed. This applies not just to the organizational and process developments, but also to the building of a trusting relationship with suppliers.

Strategic outsourcing as practiced in other industries will have to be led by top management.

Outsourcing will become the responsibility of multifunctional teams with the functions themselves concomitantly losing power.

Purchasing as a discipline will become more involved in outsourcing.

Decisions on the choice of strategic partners should primarily be based on the value they can deliver—there will be a need to decommodify activities in the drug development process.

Companies will need to develop a unified view of outsourcing, both tactical and strategic, so that suppliers can identify how they should align themselves with their customers.

Pharmaceutical companies should continue to emphasize fit and flexibility as selection criteria for partners, but should also introduce technology leadership/innovation, and, above all, customer focus and identification with end results.

Customer focus and alignment will favor the introduction of creative risk-sharing structures, together with the increased communication and trust that their implementation will require.

Strategic suppliers will become customer focused and receptive to multilayer, but clearly defined, relationships. They will seek to add value and differentiate themselves from their competition, thereby recognizing that they may limit their universe of customers (and vice versa).

Above all, strategic outsourcing requires rational decision making, which overrides the prejudices and foibles of individuals, however talented they may be.

While quality is not the only criterion in selecting partners, a continuous *mutual* quality improvement program deepens the relationship and enhances trust and confidence.

Although the study was conducted in early 1998, its findings are as relevant today as they were then. To the author's knowledge, no more recent equivalent study has been publicly reported. If the basic message from the study was that the industry would benefit from changing its practices from tactical to strategic outsourcing, then this message has not been acted upon in a significant way. To this day, few companies outsource their development strategically, although many practice strategic outsourcing in other areas of their business, such as sales manufacturing and even drug discovery.

CENTRAL SOURCING GROUP/CHIEF SOURCING OFFICER

Outsourcing has grown significantly during the past ten years across all elements of industrialized society. Companies at the forefront of this revolution are finding it necessary and advantageous to identify specific central sourcing groups of individuals at the departmental level, or, in some cases, very senior corporate staff with the responsibility for managing the outsourcing program. These outsourcing managers have, in addition to the skills and knowledge from their specific industrial background, additional "hard" and "soft" skills that allow them to deal effectively not only with individuals within their own organization but with operational and management staff within service providers.

The Outsourcing Institute, an international organization committed to enhancing clients' and suppliers' understanding of the outsourcing industry, has noted that organizing a professional staff focused on outsourcing activities (tactical or strategic) activities can bring real benefits to a company's bottom line (1998). While tactical outsourcing still predominates in many industries, including the pharmaceutical industry (where the ratio is approximately 80 percent tactical versus 20 percent strategic in R&D), a central sourcing group can serve as an important central focus for information, knowledge, coordination, and management of many activities that are already performed by suppliers but that could, by linking together into a formal alliance of professionals, form the basis of a relationship with significant value added. As part of their revolution during the past ten years, the electronics, automotive, and computer industries have established these central sourcing groups and have realized the benefit of them.

The Outsourcing Institute believes that a new breed of outsourcing manager is on the horizon—they call it a Chief Resource Officer. While individuals with this title exist infrequently within current organizational structures, directors of resources and vice presidents of outsourcing are becoming more frequently recognized within the hierarchy of firms. These corporate leaders have a complex set of skills that permit them to lead their firms to establish, manage, and deploy outsourcing principles and activities to support the bottom line. Some of these skills are:

- Experience in managing different businesses;
- Experience in managing costs;
- Ability to think out-of-the-box;
- Comfortable with change; and
- Political and cultural consciousness.

Within the pharmaceutical industry, outsourcing is not as well defined, organizationally, as it is in other industries, largely because the pharmaceutical industry has been slow to realize the benefit of an effective outsourcing program. However, in recent years, pharmaceutical R&D, responding largely to the tactical need to access external resources to meet drug delivery timetables in a fixed cost (headcount) constraint environment, has pursued outsourcing with some degree of intensity, especially within the clinical research function. Technomark Consulting Services Ltd., a specialist consultancy company to the pharmaceutical industry with particular knowledge and experience in outsourcing, recently conducted a client-funded industry survey. Their survey of seventeen pharmaceutical firms revealed the following summary information:

- Eighty-three percent of the responding companies outsource tactically, but 44 percent stated that they strategically outsource certain aspects of clinical development work. However, the strategic nature of outsourcing relationships was not a well-developed theme and did not, in most cases, emphasize strategic partnerships with contractors, as seen in other industries.

- Centralized roles/groups to manage outsourcing practices existed with only 65 percent of the respondent firms, and these roles or groups existed in the clinical organization rather than within preclinical or manufacturing.
- These central groups most often reported to the operational area with specific technical functions, as opposed to the establishment of a pan-R&D group.
- While 89 percent of the respondents stated that their group had responsibility for the outsourcing budget, there was no distinction between assigning, following, and actively managing the budget.
- Only 28 percent of the responding firms reported internal costing models but, interestingly, 50 percent stated that they compare internal versus external costs.
- While the majority of survey respondents felt that the size of the centralized role/group would increase during the next two to five years (89 percent), only 44 percent felt that outsourcing management would be a company core competence. This statement reinforces the previous approach of tactical rather than strategic outsourcing and continues to demonstrate the pharmaceutical industry's lack of motivation to move outsourcing to the next level, a level where nonpharmaceutical firms have shown outsourcing to be truly value-adding.

CONCLUSIONS

All elements of the pharmaceutical industry are experiencing pressures as a result of financial, social, political, technological, and organizational drivers. In response, consolidations have advanced in both the client and CRO segments to achieve a more effective critical mass and organizational breadth so as to provide customers with the products and services they require.

As a result of a more refined and productive drug discovery process and the downstream implications relating to a potentially significantly greater development pipeline, outsourcing of R&D capabilities, either tactically or strategically, will become a much more frequent practice. Broad-based or niche provider firms will have opportunities to expand their businesses as client firms are faced with doing significantly more work under a fixed cost constraint environment. Client firms will increasingly focus on managing external collaborations—an activity that historically has not been a core competence.

The pharmaceutical industry has always been exciting and dynamic, and it appears that the future will bring about changes in science, business, and business relationships that will continue to move the industry positively for the foreseeable future.

NOTE

Excerpted with permission from "Strategic Outsourcing in R&D" (Technomark 1998).

REFERENCES

Accenture. 1997. "Re-Inventing Drug Discovery: The Quest for Innovation and Productivity." July.

Accenture. 2000. Speed to value delivering on the quest for better medicines. March. Chicago: Accenture.

CMR International. 2000. Annual Report. London: CMR International.

Hughes, R. G. and C. E. Lumley. 1999. *Current strategies and future prospects in pharmaceutical outsourcing.* London: Technomark Consulting Services Ltd.

Ogg, M. S., et al. 2000. *Activities of the international pharmaceutical industry in 1999: Pharmaceutical investment and output.* Industry Report. September. London: CMR International.

The Outsourcing Institute. 1998. *The Source* 4(4).

PhRMA. 2000. *New medicines, new hope.* Annual Report, 2000–2001.

Technomark Consulting Services Ltd. 1998. *Strategic outsourcing in R&D.* Multiclient survey. April.

Appendix

An Overview of the Drug Development Process

DRUG DISCOVERY AND SCREENING

Innovation is a key issue for the R&D-based pharmaceutical industry and can be defined as the identification and/or synthesis of molecules not used previously to treat human diseases. The identification of potentially useful molecules is referred to as the Research or Discovery phase of R&D. In most cases, researchers identify an unmet medical need and, using knowledge about how the disease develops at the most molecular level, identify or synthesize new molecules or biological entities that will affect the disease process. This approach is known as "targeted" or "directed" research and, as our understanding of the molecular and genetic (pharmacogenomics) basis of disease evolves, it has become increasingly precise.

Traditional pharmaceutical companies have emphasized the synthesis of new molecules using organic chemistry techniques, while biotechnology companies have used biological methods for the synthesis (genetic engineering) and production (cell culture) of new drugs. Both the rate and scope of synthesis of potentially active molecules have taken a massive leap forward with the development of combinatorial chemistry, a technique that uses conventional chemistry and automated technology to produce thousands or hundreds of thousands of diverse molecules per day.

Regardless of the method used to produce a new molecule, it must be tested—screened—for activity against an appropriate receptor or target. Here, molecular biology has revolutionized the process. Historically, new molecules were tested in animal models of the disease—a time-consuming, labor-intensive process. Now they can be tested against the molecular receptor for the drug—i.e., the very molecule that must be switched on or off by interaction with the new drug to prevent or alter the disease. Advances in the technology for testing, microplate assays, and robotics now allow 400 (or possibly 10,000) potential new molecules to be rapidly screened against a target receptor in a vessel only 12.5 cm × 8.5 cm × 1.5 cm. This is known as high throughput screening. Likewise, high throughput screening can be used to test a new molecule against hundreds of receptors. The combination of high throughput screening and combinatorial chemistry provides a powerful tool to increase the chances of discovering new lead compounds. These techniques also allow random (as opposed to targeted) research, since large numbers of uncharacterized molecules (or those produced for different purposes) can be tested against many different receptors.

The structure of a lead compound is used to produce further molecules, called analogs, which are chemical modifications of the lead compound. In vitro testing identifies those molecules showing the greatest activity and these drug candidate molecules are used in the subsequent animal studies (see Figure A.1).

ANIMAL STUDIES

Once a new molecule has demonstrated potentially useful activity, animal studies are used to:

- Confirm and clarify the molecule's potential therapeutic action (primary pharmacology);
- Identify any effects other than the molecule's potential therapeutic application (secondary pharmacology);

Figure A.1   Methods of discovery research

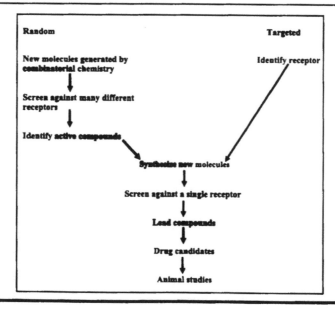

- Establish how the molecule is absorbed, distributed, metabolized, and excreted (ADME); and
- Determine whether or not the molecule shows any toxic effects in both the short and long term (toxicology).

CLINICAL DEVELOPMENT

When a molecule possessing desirable pharmacological activity in animals is found, and the secondary pharmacology, ADME profile, and toxicology results are acceptable, it may then be administered to humans. Human (i.e., clinical) evaluation goes through a number of well-defined stages:[1]

- **Phase I studies** are conducted generally in healthy volunteers. The primary aims of Phase I studies are to establish the drug's safety, that it behaves in the human body in the way that would be expected from the animal tests, that it is acceptably tolerated, and that it is bioavailable. Phase I studies are not generally designed to study the efficacy of the drug, although pharmacodynamic testing of the drug in volunteers is becoming more prevalent. Up to two hundred total healthy volunteers may be treated in a number of discrete studies. In some indications, such as cancer or AIDS, Phase I studies are conducted in patients only.
- **Phase II studies** are conducted in an intensively monitored group of patients. The objectives of these crucial studies are to establish the drug's therapeutic efficacy for the human disease and to determine the optimal safe, effective dose to treat these patients. These studies are usually performed in hospitals, but may be performed in clinics or general practices. Phase II studies usually involve groups of twenty to one hundred patients. Several different Phase II studies are usually carried out before Phase III studies can be designed and initiated.
- **Phase III studies** are conducted in large groups of several hundred to several thousand patients. The objectives are to confirm the efficacy and dose regimen, established in Phase II, in a wide range of patients and to establish that the drug had an acceptable safety profile. These studies are carried out in a setting closer to that in which the drug will eventually be prescribed, for example, in general practice.
- **Phase IV studies** are carried out once the drug has obtained marketing approval. The objectives are to widen the spectrum of treated patients (e.g., to include pediatric patients), to include variations of the disease treated (e.g., from gastritis to gastric ulcer), and to gain further understanding of drug prescribing and benefits in normal medical usage. Phase IV studies are frequently important in monitoring a drug's safety profile in a wider population.

Although there are a number of clearly defined steps in clinical development, the studies do not proceed in isolation. Thus, guidelines are laid down and enforced by law in many countries specifying the animal and genetic toxicology that must be carried out before Phase II or III studies may proceed. Depending on the drug and its application, long-term toxicology may be required before marketing approval is obtained. These studies will probably start during or at the end of Phase II, or possibly during Phase III. Figure A.2 charts the discovery and development process.

PHARMACEUTICAL DEVELOPMENT

One aspect of drug development that should not be overlooked is pharmaceutical development. Once a few milligrams of a drug has been synthesized, a team of chemists and pharmacists begins producing grams, kilograms, and potentially tons of the drug in a form that can be effectively delivered to patients (e.g., tablets, capsules, or injectable solutions). Pharmaceutical development includes chemical synthesis and scale-up, formulation development, packaging, and stability testing. Finally, if commercial quantities of a drug are to be produced and sold, the manufacturing process and the use of the drug (and its breakdown products) will have some impact on the environment. Regulators are increasingly demanding tests of potential environmental impacts.

TIME FRAMES AND COSTS

A pharmaceutical company engages in synthesis and screening efforts continuously. Table A.1 provides a range of typical times frames and costs for drug development once a drug is selected as a candidate and enters animal pharmacology.

The overall timing for a drug to move from screening to registration ranges from twenty months (a onetime record) to over fifteen years (a surprisingly frequent time

Figure A.2   Discovery and development of a new medicine

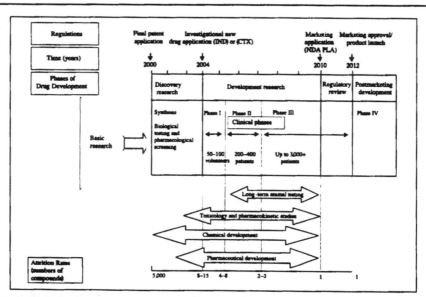

Table A.1   Time Frames and Costs

Stage of Development	Time (months)	Cost ($ million U.S.)
Animal pharmacology	6–12	1.5
Phase I	6–12	1–2.0
Phase II	6–18	2.5–10.0
Phase III	6–90	10.0–50.0
Chemical/Pharmaceutical development	up to 90	10.0–25.0
Project management	—	5.0–30.0
Regulatory toxicology	12–36	5.0–25.0
Regulatory review	12–36	0.5–5.0

(Estimates compiled from Technomark internal data. Printed with permission from Technomark Consulting Services Ltd.)

frame). The typical period is eight to twelve years, depending on a range of factors, not all of which are drug related. Two major objectives of the industry are to reduce this time to a minimum and to exploit each drug commercially for as long as possible before patent expiry and subsequent generic competition.

No published figures detail the actual cost of developing a single approved drug. However, the total cost of pharmaceutical R&D divided by the total number of new drugs produces a figure of $1 billion per drug, and this is the figure that is frequently quoted in lectures and presentations at conferences. From discovery, only 1 drug in 6,000 is ever marketed and, of these, perhaps 1 in 3 attains commercial success. The $1 billion figure takes into account all the 5,999 failures, most of which occur at the early screening stage. We estimate that for a relatively short-acting drug that is not taken chronically by patients, a true development cost of $75–100 million is feasible. More typically, costs may be $100–500 million, although this may be significantly higher for any particular drug. These figures exclude discovery costs.

NOTE

1. Clinical phases are identified according to traditional nomenclature rather than the recently adopted ICH nomenclature, which is not yet in common usage.

Contributors

Richard Ashcroft

Dr Richard Ashcroft is Lecturer in Medical Ethics at Imperial College of Science, Technology, and Medicine in London. He trained in mathematics, history, and the philosophy of science at Cambridge University, before completing a doctorate on the ethics of science, also at Cambridge. Dr Ashcroft has performed research and taught at the Universities of Liverpool and Bristol. At Bristol, he was responsible for training members of Research Ethics Committees (Institutional Review Boards) in the Southwest region of the UK. He maintains a keen interest in the ethics of research, and has published widely in the field.

Nigel J. Dent

Nigel Dent is a consultant (Country Consultancy) specializing in designing and auditing toxicology and clinical quality assurance programs for international pharmaceutical, veterinary, agrochemical, and contract research laboratories. Mr Dent has over thirty-nine years of experience within the pharmaceutical industry. Trained in applied biology and biochemistry, he has held senior management positions in clinical chemistry, toxicology, and quality assurance for Huntingdon Research Centre, Inveresk Research International, Roussel Laboratories, and Hoechst UK Pharmaceuticals.

He was a founding member of the UK QA Group, now known as The British Association of Research Quality Assurance (BARQA). He was also President of the International Society of Quality Assurance (ISQA), now reformed as The International Quality Liaison Chapter of SQA, of which he is also the President. He is President and Founding Member of the Association of Consultants to the Bioscience Industries (ACBI), a member of the U.S. SQA, the Drug Information Association (DIA), the Association of Clinical Research Professionals (ACRP), and an active member of the audit working party of the European Forum for Good Clinical Practice (EFGCP).

Roy Drucker

A graduate of the Harvard Business School's Advanced Management Program, Dr Roy Drucker manages Technomark Consulting Services Ltd., a leading management and financial consulting group. Roy Drucker qualified in Medicine from the University of Cambridge. After practicing in academia, he joined the European R&D arm of Sterling Drug, Inc. as a Medical Advisor, responsible for clinical pharmacology, drug metabolism, and bioanalysis in Europe. In 1986, he joined the Upjohn Company, subsequently Pharmacia & Upjohn Inc., where he held a range of international responsibilities. His positions there included Director of Marketing with responsibility for worldwide marketing of cardiovascular products; Executive Director, European Community Affairs and Business Systems; and Vice President, Drug Development, with global responsibility, in multiple therapeutic areas, for clinical drug development and medical support for marketed products.

Madeline J. Ducate

Madeline Ducate is Vice President of Global Site Development for Neeman Medical International, a company dedicated to providing centers of excellence through its various institutes for clinical investigation. Ms Ducate holds an MS in Molecular Biology from Old Dominion University, an MA in Management from Central Michigan University, and an MPH in Public Health and Epidemiology from Columbia University. She is intimately involved in clinical drug research, with over eighteen years working in contract research organizations in the pharmaceutical industry. After twelve years working in and teaching clinical microbiology in hospitals throughout the U.S., she joined GH Besselaar Associates, for whom she planned and designed clinical trials, and monitored investigative sites, data management activities, report preparations, and so on. She then joined PPD Development (formerly Pharmaco), where she worked for thirteen years, ultimately serving as the Vice President of Clinical Operations and Project Management for all studies conducted outside the Americas. She had oversight for two hundred clinical trials, ranging from Phase I–IV, from a single site to over 600 sites, and across more than thirty-five countries. Before joining Neeman Medical International, she worked as an independent contractor to the international drug industry, and worked with the Association Clinical Research Professionals in its certification, accreditation, and training activities.

John Glasby

Dr John Glasby is a joint Managing Director of Kendle Ely, formerly International Research Consultants, a UK-based European regulatory consultancy. In addition to managing Kendle Ely, Dr Glasby produces pharmaceutical expert reports and acts in a qualified person role to a number of companies. After obtaining a degree in Pharmacy and a PhD in Pharmaceutics, John Glasby taught pharmacy students before joining Fisons Pharmaceuticals to work in the pharmaceutical development department. Working first with prescription medicines, he then moved into development of consumer products and finally joined the registration department, eventually becoming head of International Regulatory Affairs. At that time, the major expansion of the business was in the U.S., and he spent four years running the registration department based in Boston, Massachusetts, before returning to the UK to join a small regulatory consultancy company, International Research Consultants (IRC). After some years within IRC, John Glasby was part of the management buyout team that purchased IRC from its owners, Ethical Holdings. In 1999, IRC was purchased by Kendle International, Inc., a clinical research organization based in Cincinnati.

Gillian Gregory

Gillian Gregory has a dual role at present as joint Managing Director of the UK-based European regulatory consultancy, Kendle Ely, and as European Director of Regulatory Affairs. Ms Gregory graduated from the University of Leicester in 1975 with an Honors degree in Biological Sciences, specializing in biochemistry. She began her registration career in 1976, when she joined the Registration Department of the contract research organization, Huntingdon Research Centre. She obtained regulatory experience in the pharmaceutical industry while working for E. Merck from 1978 until 1980. In 1980, she returned to Huntingdon Research Centre as Head of Registration and ran the department for eight years. In 1988, she was instrumental in the development of the UK-based regulatory consultancy, International Research Consultants (IRC), which specialized in all aspects of pharmaceutical and veterinary registration. In January 1999, IRC became part of Kendle International, Inc., a clinical research organization based in Cincinnati, Ohio. Throughout her career, she has participated in the British Institute of Regulatory Affairs (BIRA) and is currently a Fellow of the Institute.

Graham Hughes

Dr Graham Hughes is the joint founder and Scientific Director of Technomark Consulting Services Inc., a leading authority on pharmaceutical development outsourcing from preclinical to clinical and manufacturing. He is also Scientific Investment Director of Technomark Medical Ventures, a collaboration between Lloyd's Development Capital and Technomark that invests in developing biomedical companies. Dr Hughes has an MA and a PhD in inorganic chemistry from the University of Cambridge. He was awarded Research Fellowships at University of California, Los Angeles, and Boston University before joining ICI Mond Division as a Research Scientist. He then became European Development Manager for ICI's subsidiary, Atlas Chemicals in Brussels, and subsequently set up and jointly managed Effect Chemicals Development within ICI Europa. As the Regional Development Director for Zoecon, he was responsible for the development and exploitation of a range of animal health, agrochemical, and environmental health products for Europe, Africa, and the Middle East. Zoecon Europe was the prototype for the virtual company with a staff of two, conducting R&D throughout the region solely through outsourcing and collaborations. In 1987, he jointly founded Technomark Consulting Services Inc. to exploit his extensive experience with outsourcing and to build upon the market demand perceived by the company's parent, Talentmark.

Chris Keep

Chris Keep is eProcurement Development Manager with AstraZeneca, working in a global purchasing role to develop and implement AstraZeneca's B2B eProcurement strategy. Chris is a science graduate (BSc Hons, Sheffield) with over eighteen years of commercial experience in the pharmaceutical industry. He has worked in several UK sales and marketing roles, first with Merck Sharp & Dohme and then with ICI Pharmaceuticals/Zeneca Pharmaceuticals/AstraZeneca. Since 1993, he has worked as Purchasing Manager for R&D where he has worked on sourcing strategies, global supplier relationships, supplier rationalization, supplier relationship management models, and tools. He devoted about eighteen months working on the AstraZeneca merger global purchasing synergy savings project.

Patricia Lobo

Dr Patricia Lobo is the Associate Director of Clinical Operations for Technomark Consulting Services Ltd. and the current editor of *Pharmaceutical Manufacturing and Packaging Sourcer*. Dr Lobo graduated in Chemistry from the University of London and subsequently obtained an MSc in Biochemistry. She commenced her career in the pharmaceutical industry in the Quality Control Department of contract manufacturer Regent Laboratories. Moving to GD Searle, she worked in the area of Biochemical Pharmacology while gaining an external PhD for her studies on the pharmacology of plasma-binding proteins. Thereafter, she completed a one-year intensive business training program at the University of Warwick. In 1985, Dr Lobo moved into clinical research in dermatology with Stiefel Laboratories, and became a senior Clinical Research Scientist in the Oncology Group at Farmitalia Carlo Erba (UK). She also gained experience in the Marketing Department at Schering Plough (UK) as Business Development Manager, responsible for biotechnology products in oncology and virology, prior to joining Technomark in 1993. Dr Lobo has developed considerable expertise and a first-class reputation in the field of contract manufacture and packaging, providing in-depth market reports and strategic planning advice to major multi-national pharmaceutical companies.

Rakesh Nath

Dr Rakesh Nath is a cofounder and Director of DrugDev123 Ltd., the clinical research Internet portal and conference company. He completed his medical training (MBBS) at The Medical College of St. Bartholomew's Hospital, University of London. Dr Nath earned an MBA from City University Business School, and is also a Chartered Marketer (MCIM). Dr Nath has wide-ranging international clinical research, marketing, and management consultancy experience, having worked with SmithKline Beecham, Laboratoires Fournier, Besselaar/Covance, Charterhouse Clinical Research Unit, and Technomark Consulting Services Ltd. Dr Nath is a member of the Executive Committee of the British Association of Pharmaceutical Physicians, and has authored several articles and publications in the field of clinical research and strategic marketing, as well as speaking at and organizing conferences on patient recruitment–related topics.

Paul Ranson

Paul Ranson is a solicitor who has recently established PharmaLaw, a law firm dedicated to the needs of the pharmaceutical industry. He is also a consultant in healthcare law to Simmons and Simmons. Previously, he was an in-house lawyer to both SmithKline Beecham and Merck Sharp & Dohme Idea Inc. He specializes in the commercial and regulatory issues relevant to the industry. He was a member of an ethics committee for five years. Among his numerous publications, he has written or cowritten reports for PJB Publications Ltd and Informa on licensing and business development, parallel imports, and product liability.

Jeffrey S. Rudolph

Dr Rudolph possesses over thirty years of experience in domestic and international research, development, and general management of a broad range of ethical and OTC pharmaceutical products and technology. Currently, he operates his own pharmaceutical consulting service that specializes in technical solutions in the area of business and organizational development. He holds a PhD and an MS in Pharmaceutical Science from Purdue University and a BS in Pharmacy from the University of Illinois. Dr Rudolph began his career as a Senior Formulation Pharmacist with CIBA before joining McNeil Laboratories as a Group Leader and Head of the Pharmaceutical Dosage Form Pilot Plant. Dr Rudolph then joined Stuart Pharmaceuticals, where he ultimately took on the position of Director of Pharmaceutical Development, before joining Zeneca Pharmaceuticals (later AstraZeneca), where he held positions including Vice President of Pharmaceutical R&D, Vice President of the International Pharmaceutical Development Group, and Vice President of R&D Operations. Dr Rudolph is a Strategic Advisory Board Member of Vel-Quest Corporation, a member of numerous professional associations, a university guest lecturer on the drug creation process, and has presented and published papers on strategic outsourcing in drug development.

Ann Speaight

Ann Speaight is the Marketing Director of Chandos Clinical Research and a member of the Board. She trained in hospital laboratories, specializing in biochemistry and hematology, and holds a Fellowship of the Institute of Biomedical Sciences. Working with G. D. Searle, she set up and managed laboratories in London and Brussels, providing a clinical laboratory service to the independent medical sector. She joined J. S. Pathology and established a Clinical Trials Division within the routine service, offering a central laboratory facility to the pharmaceutical industry. This division ran more than 300 studies each year in all phases of drug development. Ms Speaight lectures extensively on laboratory aspects of clinical trials to members (both Study Site Coordinators and Clinical Research Associates) of the Institute of Clinical Research (ICR) and to conferences in Europe. She has written extensively on the clinical use of lab data and also on the data management aspects of studies.

Jacqui Spencer

Jacqui Spencer is Training and Development Manager for Roche Products Limited. Ms Spencer entered the pharmaceutical industry in 1989 and has held a number of varied roles within clinical data management. She began as a Data Coordinator, responsible for database design, implementation, and management, and advanced to European Clinical Data Manager for her first company, then moved away from the pharmaceutical company environment and into clinical data management consultancy for Technomark Consulting Services Ltd., based in London. While at Technomark, she helped pharmaceutical companies identify appropriate CROs to use for data management activities—either for a specific project or for all such activities. Additionally, she was involved in evaluating the data management market in general and, toward the end of her five years with the company, she was involved in evaluating new data management systems, electronic data capture, and the use of the Internet for clinical data management. In November 1999, she joined Roche Products as the Training and Development Manager for the clinical data management group.

Lucien Steru

Dr Lucien Steru works as a consultant in Belgium. He graduated in medicine (1979), psychiatry (1981), and neuropsychopharmacology (1983) from the Faculty of Medicine Pitié-Salpêtrière in Paris, France. He performed his military service in behavioral psychopharmacology research in the French Air Force in 1980. While he was Assistant-Professor in Pharmacology at Faculty of Medicine Pitié-Salpêtrière (1977–1983), he practiced psychiatry as a resident and conducted research in behavioral pharmacology, mainly in the area of depression. He founded the first French modern Contract Research Organization in Phase II–IV clinical research, I.T.E.M. (1982). He also founded the first Contract Lab in behavioral psychopharmacology (I.T.E.M.-Labo, 1984). Dr Steru owned and managed I.T.E.M., which was a member of Verum (the largest European Clinical Research Alliance founded in 1992), from 1982 to 1997. It had reached 250 employees in Europe when it merged with Phoenix International, a Canadian public Contract Research Organization, in 1997. Dr Steru became President, Chief Operating Officer, and Board member of Phoenix International. He developed the European activities in seven areas of service with over 700 employees. He retired in 2000 when Phoenix International was purchased, although he continues to work as a consultant.

Joseph S. Tempio

Dr Joseph S. Tempio is the Director of the Pharmaceutical Technical Services practice of Tunnell Consulting, a thirty-nine-year-old, employee-owned, diversified consulting practice that delivers technical and regulatory compliance consulting services to the pharmaceutical and healthcare industry. Dr. Tempio is responsible for the deployment of more than fifty consultants working in the practice, and is a company Vice President and member of the Board of Directors. Prior to joining Tunnell, he was at IBAH Pharmaceutics Services (now Omnicare Clinical Services), where he served as Vice

President of Operations and later as President. Before IBAH, he was Director of Scientific Affairs and ran product development for Lederle Consumer Healthcare (now a division of American Home Products). For the previous thirteen years, he held several R&D management positions in the consumer and pharmaceutical divisions of SmithKline Beecham. He holds a BS in Pharmacy from Duquesne University, and an MS and PhD in Pharmaceutical Sciences from Rutgers University.

Nadia Turner

Nadia Turner is a pharmacologist by training (BSc Hons, Leeds University) with over fourteen years of experience in the pharmaceutical industry. She has worked in both preclinical and clinical research (with Boehringer Ingelheim and Procter & Gamble Pharmaceuticals), as a consultant specializing in outsourcing (Technomark Consulting Services Ltd.), and as Development Strategic Sourcing Manager for Zeneca Pharmaceuticals. Following a brief spell as a Global Project Manager, Nadia is currently Associate Director, Global Development Sourcing, at AstraZeneca Pharmaceuticals, where she leads a cross-functional team in developing and promoting sourcing best-practice decision making, implementation, and evaluation across AstraZeneca's global drug development communities.

Tim Wright

Tim Wright is Vice President of corporate sourcing at BTG plc, where his principal responsibility is acquiring drug and technology candidates from multinational corporations based in the UK and Europe. Prior to this, he held roles as Director of business development for DevCo Pharmaceuticals Ltd., a virtual company focusing on neuroscience drug development, and Glaxo Wellcome Research and Development, where he was head of worldwide medical resourcing and contracts. Tim has a PhD in pharmacology from London University and an MBA from London Business School. He has chaired the European Pharmaceutical Contract Managers Group and has often presented or chaired conferences on pharmaceutical outsourcing, business development, and risk sharing with development partners.

Index

Milton Keynes UK
Ingram Content Group UK Ltd.
UKHW052018071024
449327UK00027B/2320